Extending and Improving Your Home

Extending and Improving Your Home

An Introduction

M.J. Billington BSc, MRICS, Consultant

Managing Director
Construction Auditing Services Ltd

C. Gibbs BSc

Building Surveyor and Consultant

WILEY-BLACKWELL

A John Wiley & Sons, Ltd., Publication

Blackwell Publishing was acquired by John Wiley & Sons in February 2007.
Blackwell's publishing program has been merged with Wiley's global Scientific,
Technical and Medical business to form Wiley-Blackwell.

Registered office: John Wiley & Sons, Ltd, The Atrium, Southern Gate, Chichester, West Sussex,
PO19 8SQ, UK

Editorial offices: 9600 Garsington Road, Oxford, OX4 2DQ, UK
  The Atrium, Southern Gate, Chichester, West Sussex, PO19 8SQ, UK
  2121 State Avenue, Ames, Iowa 50014-8300, USA

For details of our global editorial offices, for customer services and for information about
how to apply for permission to reuse the copyright material in this book please see our
website at www.wiley.com/wiley-blackwell.

Library of Congress Cataloging-in-Publication Data

Billington, M. J. (Michael J.)
   Extending and improving your home : an introduction / M.J. Billington,
Clive Gibbs.
      p. cm.
   Includes bibliographical references and index.
   ISBN 978-1-4051-9811-0 (pbk.)
1. Dwellings–Remodeling. 2.   Building–Superintendence.   I. Gibbs, Clive.
II. Title.
   TH4816.B55 2012
   690'.24–dc23

                                                          2011053110

A catalogue record for this book is available from the British Library.

Wiley also publishes its books in a variety of electronic formats. Some content that appears
in print may not be available in electronic books.

Cover design by Sandra Heath
Cover image courtesy of iStockPhoto and Shutterstock

Set in 10/12.5pt Minion by Thomson Digital, Noida, India
Printed and bound in Singapore by Markono Print Media Pte Ltd

[1   2012]

Contents

Preface

In April 2008, I (MJB) was invited to provide support to the 'Ask the Expert' stand at the Ideal Home Show at Earl's Court in London. Being a Chartered Building Surveyor and Building Regulations expert, I was anticipating lots of questions from the public on defects in houses and on specific regulation queries. These occurred as expected but were far outweighed by the multitude of questions on altering and extending homes. It quickly became clear that because of the recession we were in a new era whereby, instead of moving house, people were now far more interested in staying put and improving. It also became clear that most people did not have the foggiest idea about what they could do and how they could go about doing it. People even brought in rough floor plans of their houses and I ended up doing off-the-cuff sketch schemes for everything from fitting a loo to designing an extension!

Thus was born the idea for this book. It is entitled *Extending and Improving Your Home: An Introduction*. The last two words are important because even in a book dedicated entirely to this subject, it is impossible to cover everything comprehensively. What we have done is provided a focused guide to all the areas in the field of home improvement that cause most concern for householders, and at the end of each chapter we provide sources of further information in terms of written texts and web sites. It should be noted that the sources of information provided are for information only. The authors have used these sources in the past and found them to be accurate, informative and reliable. We can make no promises that they will continue to be so in the future. Therefore, their inclusion in this book must not be taken as a recommendation or endorsement and we are, of course, not responsible for their content.

So who should purchase this book? Although it is aimed at anyone considering improving or extending a house and is written in non-technical language as far as this is possible, the technical sections can be applied to the construction of brand new homes as well as extensions. Therefore, it will be of use not only to householders but also to small builders, tradesmen such as bricklayers and carpenters, students on the first year of construction and architectural design courses at tertiary and higher education establishments and building professionals such as estate agents, building surveyors, architects and other designers.

Our aims in writing this book are to

- Guide people through the maze of legislation that affects most building alterations and extensions;
- Show how to go about getting a project realised in terms of design, choice of contractor and construction, successfully and to budget;
- Give sound technical solutions for all the elements of a project that also comply with Building Regulations.

Combining over 70 years of experience specifically in the field of home improvements and alterations, we also point out the common pitfalls that can occur and show how they can be avoided.

More specifically, in Chapter 1, the book starts by considering the pros and cons of moving house as opposed to extending or altering what you have already. It discusses the rise of DIY and considers the implications of getting someone else to do it for you. The processes and procedures of home alterations and extensions are discussed in detail and the chapter ends with a series of tips to assist you in the design and construction of your project.

Chapter 2 goes into detail on the labyrinth of legislation that applies to this sector of construction and shows what you need to comply with and how compliance can be achieved.

In Chapter 3, we look at the various ways you can arrange and organise the project to ensure that you realise your ideas successfully and do not get ripped off. And if things still go wrong, we suggest methods for ensuring recovery of the situation.

Chapter 4 looks at those issues that affect the property as a whole no matter what you intend to do, such as structural stability, weatherproofing and maintenance.

The first, general part of the book ends with Chapter 5. Here, we consider the importance of gathering information on the interior of your house and on the site outside to avoid any nasty surprises arising when the work starts.

The technical part of the book covers Chapters 6–12 and discusses each element of the work (foundations, floors, external and internal walls, roofs, external and internal finishes and services) and includes detailed examination of the Building Regulations that apply in each case.

The last part of the book provides practical guidance on specific projects. Thus, Chapter 13 looks at all the ways you can improve the thermal efficiency of your home (highly topical at present)and Chapter 14 looks at specific conversions to increase the space within your home without recourse to extending such as, loft, garage and basement conversions.

Finally, we provide two case studies. Case Study 1 demonstrates the economic advantages of improving the thermal efficiency of your home by reference to simplified calculations for a number of common upgrades.

Case Study 2 shows a typical loft conversion and points out the most important regulatory and technical considerations in what is a surprisingly difficult area of work.

It must be stressed that this book is a guide to commonly accepted construction details, techniques and methods of building. Since the authors cannot be aware of all the circumstances of a particular case, they cannot take responsibility for inappropriate use of the guidance contained in this book. Additionally, the information given on the Building Regulations throughout this book is provided as a guide to the regulations, not a substitute for them. When using an Approved Document, it should be remembered that the guidance in them is not mandatory and differences of opinion can quite legitimately exist between controllers and builders or designers as to whether a particular detail in a building design does actually satisfy the mandatory functional requirements of the Building Regulations.

We hope that by using this book, you will be able to make decisions on the basis of facts rather than feelings, hearsay or rumour. We will feel that we have done our job in writing this book if after reading it you are

■ Better informed,
■ Aware of the importance of forward planning and
■ Forewarned and, therefore, forearmed.

The law is stated on the basis of cases and other material available to us on 1 September 2011.

M. J. Billington
Clive Gibbs

1 Extending and improving your home – an introduction

Houses are like people. They are conceived (in the mind of the designer). Over several months they grow (during construction) and they are eventually born (when completed). They grow and have to change to meet the changing circumstances of a growing family (extra bedrooms or new conservatory) and gradually mature as they grow older. If they go out of fashion they are updated with all the latest trends (en-suite bathrooms and central heating). Sometimes they get sick (dry rot, woodworm and rising damp) and have to be cured so that their lives can be extended. Eventually they get very old and then it may be necessary to put them out of their misery (although we do not advocate euthanasia here!).

Unlike people, houses are virtually immortal and as they get very old they usually become more interesting and more loved (even if they are a bit crotchety).

There is no doubt that carrying out significant changes to your own house can be incredibly rewarding. We all have our own ideas about the sort of house we want to live in but, unfortunately, we nearly always have to buy a house that does not come up to our expectations. So then we have a choice, keep moving in the forlorn hope that we will find the perfect house, or try to make the one we are living in as near as possible to our perfect house.

It is reckoned that the three most stressful things that can happen to you in life are (in order of greatest stress)

1. The death of a loved one,
2. Divorce and
3. Moving house.

In this book we cannot help with the first two but, hopefully, we can eliminate the stress of the third by helping you to avoid moving.

In this chapter, a few basic questions are covered, such as

- Why would you want to extend or improve your house anyway?
- Where did this idea of controlling or doing the alterations yourself come from?
- What processes and procedures do you need to go through to have the most chance of success when making these changes?
- Are there a few simple things that you need to know to help you avoid the tiger traps that you might otherwise fall into?

Extending and Improving Your Home: An Introduction, First Edition. M.J. Billington and C. Gibbs.
© 2012 M. J. Billington and C. Gibbs. Published 2012 by Blackwell Publishing Ltd.

WHY EXTEND OR IMPROVE YOUR HOME?

There are almost as many answers to this question as there are ways of actually doing the work. It is incredible to think that only 5 years ago property prices seemed that they would continue inexorably to rise by considerably more than the rate of inflation year after year. After all, it had been like this for such a long time that your house was seen (perhaps incorrectly) as not just a place to live but also as an investment. How else, the theory went, could you expect to beat inflation so easily. People tended to move rather than face the stress and strain of '*having the builders in*' and they were often put off carrying out improvements for fear of getting ripped off by unscrupulous tradesmen and builders, and then of course there was '*the mess*'! Strangely enough, when they actually did move, they would normally replace the bathroom and kitchen and carry out a complete redecoration!

How things have changed. The credit crisis of 2007, the subsequent recession and the resultant fall in houses prices (and current stagnation in the market) have convinced most people that unless you have to move, do not – improve or extend instead.

According to a report by Sainsbury's Finance, an estimated £3.2 billion worth of personal loans was taken out in 2010 for home improvements alone, and it is likely to be a similar figure in 2011 – with one in five personal loans being taken out solely to pay towards improving people's homes.

Arguments against moving include

- **The high cost**: At least £7500 in stamp duty for a property over £250 000, professional fees payable to estate agents and solicitors amounting to another 3 or 4% of the sale price, removals costs adding another £1000 or £2000 – for an average priced house it can cost between £15 000 and £20 000 to move, and much more as you go up market.
- **The stress of selling your house and buying another one, especially in a stagnant market**: According to Hometrack (www.hometrack.co.uk) the average time it takes to sell a home stood at 9.4 weeks in July 2011, which is almost the longest time taken to sell a property since the survey began in 2001 and '*despite weak consumer sentiment the housing market is currently in broad equilibrium although prices continue to slowly edge lower*', in other words – stagnant!
- **The difficulty in raising a mortgage deposit**: The deposit needed will vary with the type of mortgage sought and, generally speaking, the lower the deposit, the higher the interest rate you will have to pay and the higher the arrangement fee. At present minimum deposits start at 10% of the valuation and go up depending on the product sought. That is a sum of at least £25 000 to be raised.
- **The general state of the economy**: Do you really want to spend all your savings and end up with higher mortgage repayments when you may not have a job next week?

So, unless you have to move for personal or business reasons, why spend so much money moving when you could spend it on the house you currently live in and make it how you want it to be?

Some of the reasons often put forward for extending and/or altering your home are

- **It will add value to your home**: To check if this is true you should first find out the current market value of your home by getting it valued by an estate agent. Then get an estimate of its potential value once the proposed improvements have been completed (not forgetting to include the estimated cost of the improvements in your calculation). Unless your home is in an exceptionally sought after area and/or is in need of drastic improvements and repairs it is unlikely to gain a great deal in value by being improved. Additionally, the authors have seen cases of 'over improvement', where a house has been priced out of its area by too many extensions and alterations.

- **Improvements will make your home more attractive to potential buyers**: This is undoubtedly true, provided that the improvements are well executed, well designed and are in keeping with the design of the house and its general location. If poorly constructed and out of sync with the ambience of the area the 'improvements' will actually make your house less attractive to buyers. It is often the case that, having gone through the processes of alteration and improvement, many people decide that they actually rather like their house, so they decide not to move after all! Additionally, you could argue that if you spend, say, £25 000 making your house more attractive to buyers, why not simply reduce the price by £25 000 and save yourself the hassle?
- **Construction companies may well be lowering their prices to compete for work to stay afloat**: Do not ever use this as an excuse to improve your house. Whilst it is true that general construction costs are relatively low at present, companies that undercut the general market rates to get work usually end up cutting corners in the construction, or they look for ways to claim extras during the contract works, or they go bust halfway through the project! Be prepared to pay the market rate using reputable builders, after all if you pay peanuts you get monkeys (and you do not want to be the monkey!).
- **The improvements will make my house a better and more comfortable place in which to live**: The best reason of all and the only one that really matters in the end.

Ideas for extending and improving

Home improvements can be divided into the following general classifications:

- Ordinary regular maintenance, such as painting exterior woodwork (windows, doors, fascia boards, rainwater gutters and downpipes etc.), replacing the odd slipped roof tile and changing tap washers.
- Necessary repairs brought about by lack of maintenance, accidents or emergencies, such as replacing perished rainwater gutters, downpipes and fascia boards, renewing roof coverings and unblocking foul drains.
- Repairs needed because of defects that arise, such as renewing a defective damp-proof course or treating timbers for woodworm or rot.
- Internal redecoration to reflect personal tastes.
- Internal structural remodelling by removing walls and chimney breasts to create more or different space.
- Upgrading out-of-date services installations, such as rewiring and new central heating systems.
- Refitting and updating kitchens, bathrooms, utility rooms and bedrooms (e.g. installing fitted furniture).
- Installing new bathrooms, kitchens and utility rooms within the existing spaces (e.g. converting a bedroom to an extra bathroom after the kids leave home!).

- Converting 'dead' space to living space, such as loft, garage and basement conversions and the conversion of outhouses and sheds to, for example, garden offices for people working from home.
- Single and two storey extensions to provide additional improved accommodation of any kind.
- Upgrading the thermal performance of the dwelling to make it more energy efficient and reduce household energy bills, such as new windows and external doors, upgraded insulation in walls, floors and roof and the installation of energy efficient heating and lighting systems.
- Installing energy efficiency measures, such as photovoltaic panels, solar water heating, wind turbines and so on.

It is quite a long list and when people first climb onto the property ladder they are usually unprepared for the likely expenditure involved in owning a home of their own. In Chapter 4 maintenance issues are looked at. It is made clear that regular planned maintenance, not responsive maintenance, is the best and most cost effective way of looking after your home. The old adage of 'a stitch in time' really is the best answer when it comes to looking after your house.

DO-IT-YOURSELF

In Chapter 3, the pros and cons of DIY against getting a builder to do the work are considered. The DIY phenomenon is, in fact, quite a recent innovation and saw the start of a move away from the traditional way of commissioning work from a builder. In his poem, *A Truthful Song* written in 1910, Rudyard Kipling writes in a section on The Bricklayer:

> '*I tell this tale, which is strictly true,*
> *Just by way of convincing you*
> *How very little, since things were made,*
> *Things have altered in the building trade.*'

This attitude was true up until the end of World War II. It is generally agreed that the DIY movement (like so many other things) started in the United States in the late 1940s and early 1950s. In the immediate post-war period, as the American economy boomed and full employment returned, a comprehensive building programme spread private dwellings across much of the United States. After the upheaval and disruption caused by World War II, it is as if many citizens, including returning ex-servicemen, sought the safety and security of their own homes, and improving and extending their property was an expression of this need. This desire was assisted by two major technological developments:

1. New materials, such as hardboard, plasterboard, emulsion paint, Formica and plastics.
2. In 1946, the American tool company Black & Decker introduced its Home Utility line, the world's first popularly priced drills and accessories. Such was the popularity of these products that the one-millionth $\frac{1}{4}$-inch Home Utility drill came off the assembly line only 4 years later in 1950.

DIY took a little longer to reach the United Kingdom. This is mainly due to the following stark differences that existed between Britain and the United States as a consequence of the war.

- Many UK cities were partially destroyed by mass bombing raids and strikes from unmanned explosive devices (the V1 flying bomb or 'doodlebug' and the world's first ballistic missile, the V2 rocket) throughout the war. In London alone, over one million houses were destroyed or damaged during the Blitz between 7 September 1940 and 10 May 1941. By contrast, there were no air raids on the American mainland during the war and no buildings were destroyed.
- The housing stock in Britain was much more diverse and, of course, covered a much longer period.
- At the end of the war the United Kingdom was effectively bankrupt. During the war the United States had supplied war material to Britain and the other Allies under a program known as Lend-Lease. A total of $31.4 billion worth of supplies and equipment was supplied to Britain. Although there was no charge for the Lend-Lease aid delivered during the war, in 1946 the post-war Anglo-American loan further indebted Britain to the United States, resulting in an initial loan value of £1.075 billion. Payment was to be stretched out over 50 annual payments, starting in 1951. The final payment of $83.3 million (£42.5 million) was made on 29 December 2006.
- The level of owner-occupation in Britain in 1951 stood at a mere 28%. Only 18% was local authority rented housing, so the majority (54%) was owned by private landlords.

So, in Britain, returning ex-servicemen often found that their homes had been destroyed, the economy was in severe recession so jobs were hard to come by, inflation was at around 5% (historically high at that time) and like it or not they would be renting substandard housing from a private landlord. Consequently, they had no incentive or interest in improving their homes.

Recovery was slow and the shortage of new houses in Britain until the late 1950s meant that buyers had to take what they could get, which often meant older properties. Traditionally, such older properties were decorated in dark and sombre colours, few had inside toilets or bathrooms and none had any form of central heating. However, the mood eventually changed and during the latter part of the 1950s the economy improved, resulting in a housing boom which saw the numbers change drastically, so that by 1961 the number of owner-occupiers had increased to 43%, occupation of local authority rented housing had increased to 24% and the private rented sector had shrunk to 33%. People in older properties wanted to upgrade and brighten up their houses but the shortage of money and the lack of suitable contractors meant that they had to do it themselves.

Two principle sources of information came to their rescue:

1. Magazines such as *Practical Householder* (first published in 1955 and the first DIY publication of its kind) and *Do It Yourself* magazine (first published 1957 and said to have reached a readership of around 3 750 000 monthly by March 1960).
2. The launch of the first DIY series of programmes on British television in 1955 (Bucknell's Do It Yourself), which attracted 7 million viewers. This was followed by a further series in 1962 'Bucknell's House', which followed a 39-week BBC project renovating a house, bought for £2250, in Ealing (and now reckoned to be worth £800 000!).

It is claimed that these programmes alone kick-started the DIY industry in the United Kingdom, which today is worth around £8 billion annually. However, in the 1960s DIY was a hobby of the few and obtaining the necessary materials in the late 1950s and early 1960s was no small task. These had to be obtained from traditional builder's merchants and timber yards (where the service could be intimidating to the average DIY enthusiast) or small 'handyman' shops, often in backstreets. Therefore, it may be no coincidence that in 1969 Britain's first major DIY store was founded by Richard Block and David Quayle; the first store opened in Portswood, Southampton, in a disused cinema. Although initially called Block and Quayle, the name was soon shortened to B&Q.

Until recently the DIY industry was dominated by the big four – B&Q, Sainsbury's Homebase and Wickes, although Focus ceased trading in May 2011. Recent consumer research indicates that the superstores are dominant in the sector but it is equally clear that there is a substantial amount of business that goes to non-specialists. For example, Wilkinsons is highly successful and attracted more customers than Focus.

Current research has shown that there has been a decline in DIY sales in the last 2 or 3 years; some blame it on the recession but the causes may be more complex. One of the major developments of the latter stages of the consumer boom up to late 2006 was that people tended to stop doing DIY and started employing others to do it for them. As tradesmen are more likely to use a builder's merchant, DIY retailers lost share. At present there seems no doubt that the DIY industry will stagnate until a general improvement in the economy. It is evident that people tend to do most DIY soon after moving house. With far fewer transactions in the current housing market, this alone means that there will be less cause to do DIY.

EXTENDING AND IMPROVING YOUR HOME – PRACTICES AND PROCEDURES

To get a good understanding of the basic procedures behind the improvement and extension of your home, the most comprehensive information may be found on the Royal Institute of British Architects' (RIBA) web site (http://www.architecture. com). Known as the *Outline Plan of Work* it takes you through a number of stages from the inception of the project, when you start having vague ideas about what you want to do, to the post-completion stage, when you tie up all the loose ends after the project has come to a satisfactory conclusion. There are several stages identified by the RIBA.

Stage A – Appraisal

Here you identify what your needs are and what you want to achieve. You would also work out the financial practicalities of a range of options that you might have and look at any possible constraints on any of the proposals. After carrying out these processes you should be in a position to list out under each of your options the pros and cons, so that you can decide which of them (if any) you should proceed with. Such processes are usually called 'feasibility studies'.

Stage B – Design brief

From your feasibility studies you will come up with an initial statement of requirements. This needs to be developed into a Design Brief, which should confirm the key requirements and constraints. You should also try to identify how you intend to get the work done (e.g. employ a professional designer to do everything for you, employ a builder to do all the work, buy the materials yourself and employ subcontractors to do each stage or do it all yourself etc.) and you will also need to decide on the experts you need to employ to design those parts of the works that you cannot do yourself, such as a structural engineer or specialist services engineer.

Stage C – Concept

You have now made some positive decisions about what you want to achieve and how you want to achieve it. So now you need to develop your design brief into what architects call a Concept Design. At the end of this stage you should be able to determine the form in which the project is to proceed and ensure that it is feasible functionally, technically and financially. You would also review the decisions you have made about how you want to get the work done and what it is likely to cost.

Stage D – Development

Your concept design can now be worked up into plans, specification of principle materials (external walls, roof finish, position and type of windows and external doors), the structural support system and the major services, and the costings can be checked to make sure you are still within your budget. This stage ends with the preparation and submission of your planning application (if you need one).

Stage E – Technical design

Now you get down to the nitty gritty of the technical detailing. This is where you really do need expert advice so that you can be confident that your design will

- Pass building regulations;
- Conform to acceptable standards of building construction;
- Be sufficiently well detailed and described to allow contractors who are tendering to do this accurately and
- Not contain any design elements that will be unsafe to construct.

Stage F – Production information

Production information is needed so that you can obtain accurate competitive tenders if you are going down this route to have your work done. It usually means preparing a specification that can be priced by the various contractors who are tendering for the work. The specification will vary in detail depending on the complexity of the work; for simple schemes it could be simply a written description of what you want to do, structured so that each item of work is described and can be priced separately. This is termed an itemised specification and it is normal to describe the pieces of work by trade and/or element, such as foundations, brickwork, joinery, plumbing, roof tiling and so on. The advantage of an itemised specification is that when it comes to doing the work and you decide to change your mind about, for example, the number of socket outlets you would like, then that element can be readily identified and a revised cost worked out accurately. If you just have a single figure for the entire project, how do you know whether or not you are being overcharged for the item. Of course, if you employ a professional to manage the job for you he will be able to advise you and negotiate with the contractor from a position of knowledge and experience.

For very small jobs it would not be necessary to prepare a fully itemised specification but it is still a good idea to have an accurate description (sometimes called a 'Schedule of Work') of what you want to achieve so that there are no misunderstandings with the contractor.

This stage ends with the preparation and submission of your building regulations application and planning application (if you need one).

Stage G – Tender documentation and stage H – tender action

If you have decided to go out to competitive tender for your work (it is normal to obtain at least three separate tenders) you will need to prepare three separate sets of all the design and specification information and tell the contractors the date when you want the tenders to be returned. For larger contracts, late submission of a tender would normally disqualify the tenderer, however for smaller scale domestic work it is usual to accept the tenders even if they are a bit late, and in boom times it is not unknown for tenderers to not bother to submit a price! You should, of course, gauge their interest in submitting a price before sending them the documents. Stage H consists of evaluating the tenders when they have been returned. Do not just look at the bottom line and do not imagine that you have to accept the lowest tender. If you have asked for itemised pricing you can compare individual prices for the different parts of the work and see if a contractor has loaded a particular item because he thinks that it may lead to extra work that can inflate the costs later in the contract. One such item is foundation excavation. If, when the building control officer arrives on site, he is not happy with the depth of the foundation excavation because the subsoil is not as expected, he may ask for the foundations to be dug deeper. Such unknowns are usually dealt with by including a 'provisional sum' in the specification where you can tell the contractor to allow extra money to cover this eventuality. If the foundations do not have to be increased in depth then you simply disregard the provisional sum. Another way to budget for unexpected items is to include a 'contingency' sum in your specification. This is usually done by stating that a certain percentage (say 10%) is to be added to the final tender figure for works as yet unknown or undefined and you may or may not need to spend this.

Stage J – Mobilisation

After you have received the tenders and have decided which one to accept you will need to enter into a contractual arrangement with the successful contractor. As is discussed in Chapter 3, you can opt for a formal standard contract or can simply accept the contractor's price by a letter of confirmation. Most reputable contractors will have standard terms and conditions of business and you should study these carefully before signing the agreement. They will tell you what the responsibilities of each party to the contract are and will explain how payments are to be made to the contractor. At this stage you should also agree how information will be given to the contractor and discuss how the works will take place. This will include the start on site date and programme for the works and items such as protection of your premises, areas allocated to the contractor for storage of materials, temporary water and electricity supplies for the works, toilet accommodation for use by the contractor and the hours when the contractor will be permitted to work. You should also check that they have the correct insurances in place.

Stage K – Construction to practical completion

At last you have reached the construction stage of the contract and the works are underway. This is where you have to decide whether to control the works yourself or employ a project manager to look after the job, or allow the contractor to organise everything. Much will depend on your own knowledge and abilities (including how much spare time you have) and the complexity of the job. All these issues are discussed in more detail in Chapter 3.

Stage L – Post-practical completion

When a contract nears completion a number of administrative issues need to be dealt with. These include the final inspection by building control and the final payment to the contractor. On larger contracts it is usual to hold back a small amount of money, say 2.5–5%, of the contract sum for a period of 6 months during the defects liability period so that money is available to put right minor defects in case the contractor refuses to do so. On most small domestic scale jobs this will not normally apply; you should check the terms and conditions agreed with the builder at the start of the contract, where it will usually state that the final payment must be paid at practical completion. Practical completion is usually deemed to be when you take over and use the completed works, although in some cases the practical completion may be phased so that certain parts of the works are taken over before the completion of others. You would only pay for the parts as they were completed of course. The contractor will arrange for any necessary tests of services to be carried out and will arrange the issue of the relevant building control certificates where Competent Persons have been used (for electrical installations and the installation of replacement windows and external doors, for example). You should also obtain commissioning certificates for new boiler installations or fixed fuel burning heating appliances and full operating and maintenance instructions. Finally, just before the work is complete and ready to be handed over, you should go round the job with the contractor and make a comprehensive list of anything that is not to your satisfaction within the bounds of the contract. This stage is called snagging and the final payment should not be made until all the snagging items have been completed.

PRACTICAL DESIGN TIPS

As stated at the beginning of this Chapter, carrying out significant changes to your own house can be incredibly rewarding. If you have not done it before it can also be a daunting experience. A few simple things that will help you avoid some of the major pitfalls to which you might otherwise succumb are listed here. However, if you are in doubt never be frightened to seek expert advice. It will save you money and stress in the long run.

Obtaining planning permission can be sometimes simple, sometimes very complex and this process can be influenced by numerous factors, such as

- **The materials and style of design**: It is very often the case that people wishing to extend their property want it to blend in or be 'sympathetic' with the original building. On the other hand there are those who wish to be more adventurous, perhaps using lots of glass and non-traditional materials. There is no hard and fast rule that will guarantee planning permission is obtained. The common statement that extensions should be in keeping with the character and appearance of the local street scene is very true, if you live in a street where all properties are very similar. However, if you live in an old detached stone built property with an interesting history it could be more sympathetic to the original property to have a more contemporary design, thereby making clear what is old and what is new. This will retain the historical character and appearance of the original building and may be a more agreeable option for the planners.
- **Other planning legislation applying to the property**: The type of materials and style of improvement could be influenced by different types of legislation. For example, Conservation Area Consent, National Parks legislation, the situation of the property in an Area of Outstanding Natural Beauty, Listed Building Consent and so on.
- **Natural light and privacy**: This is a case of being considerate to neighbours. Even if you submit a design that ticks all other boxes, it could be refused planning permission if it would adversely affect your neighbour's use and enjoyment of their property. A common complaint is that your extension would greatly reduce the amount of light entering the neighbour's property. A basic rule is to draw a 45° line from the centre of your neighbour's window to your extension, if you encroach on this line there may be cause for concern. Extensions and improvements that may overlook a neighbour's garden or into their house could also be refused. This could apply to single storey extensions with accessible flat roofs, for example. If there is a risk of directly opposing windows, then a distance of 21 m is required.
- **Scale of proposal**: Extensions should be subservient to the original house and, therefore, should not dominate it. Proposals that extend the full height, length or width would not normally be acceptable.
- **Style**: The architectural style of the property should be consistent also. For example, roof, window and door styles and sizes should be consistent.

Many local authorities produce householder guides to domestic alterations and extensions that consider design factors for different situations. Additional guidance can also be obtained online from the Planning Portal (www.planningportal.gov.uk).

Described so far are some of the design principles that apply to the project conceptually. At a more basic level you might also like to consider the following design tips. These have been picked up over a lifetime spent designing and building alterations and extensions to buildings. They are offered so that some of the pitfalls that have befallen others can be avoided.

- **Connecting the extension to the house**: To most people, the logical way of connecting the extension is as shown in Figure 1.1. Here the extension side wall lines up with the

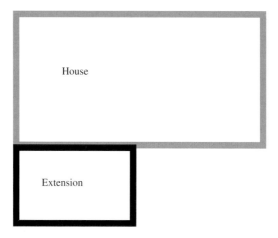

Figure 1.1 Positioning of extension – in alignment with house.

side wall of the house. Now consider Figure 1.2. The extension is offset so that it is no longer in alignment with the side wall of the house. Why would this be done? Mainly because it disguises the junction between the house and extension. If the aim is for the extension to merge with the house and not stand out like a sore thumb, then it is necessary to consider the wall and roof materials. Most houses in the United Kingdom have walls finished externally in facing bricks. Over many years the design of bricks has changed and the most noticeable change is in the size of the bricks. A modern facing brick has a height of about 65 mm. A brick produced in the housing boom between the two World Wars had a height of about 75 mm and Victorian bricks were often larger than this. So unless you have a fairly modern house you will not be able match the height of the brick courses, and you would not be able to match the texture and colour of the bricks either. If you exactly line up the extension with the house, the resulting

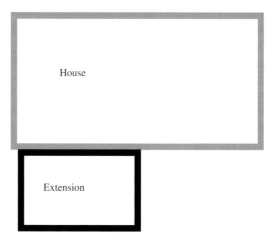

Figure 1.2 Positioning of extension – offset in relation to house.

junction will be difficult to build and will look awful. Merely offsetting the extension by the length of a brick solves both the constructional and aesthetic problems. It is also stronger structurally as it moves the work of cutting into the existing building away from the corner.

- **Designing the roof**: Generally speaking, if a house has a pitched roof the extension will look better if it also has a pitched roof, preferably of the same pitch. The same applies for flat roofs. There will be the same problems of matching materials for tiled or slated pitched roofs that there are with matching bricks. If an extension roof merges with the house roof (for single storey extensions to bungalows and two storey extensions to two storey houses) it may be possible to use the tiles or slates taken off the existing roof at the point of connection to put these on the most exposed slopes of the extension roof. The new tiles can be put on the slopes that are less obvious or hidden from view.

- **Design tips inside the extension**: Try to think in three dimensions. It is not uncommon for college students on the first year of their design courses to draw the ground and first floors of a building without adequate thought for the continuity between the two different plans. In such cases the staircase may not work out or they find that the upstairs toilet can only be drained via a pipe that comes down through the middle of the lounge! The golden rule is to plan ahead. Do not just plan alterations for the immediate future but try to think about how the alterations might be adapted should your circumstances change. Have a definite plan for the future development of your house and do not let your current alterations jeopardise future plans. An obvious example is the construction of a single storey extension on a two storey house. Is it possible that you may want to extend upwards on top of this in a few years time? It is not just a matter of making sure that the foundations can take two storeys. You must also consider the position of the extension, not only in relation to the ground floor but also in relation to the first floor. And you may not want a flat roof on your extension but it is easier to turn a flat roof into a future first floor than it is to do the same to a pitched roof!

- **Walls**: It is often a good idea to construct the internal walls of an extension in the same type of materials as the external walls. For example, if your extension has cavity masonry external walls then the inner leaf will be in lightweight concrete blockwork, so build the internal partition walls in the same type of blocks where possible. The temptation is to build the internal walls in plasterboard-lined timber stud partitioning. The reason for matching the materials is that if you do not, a crack will form at the junction of the studwork and the blockwork as the two materials move differently. This can be filled when redecorating the extension at some later date but in the meantime the crack will be evident. If you must build the internal walls in a different material it makes sense to dry line the external walls instead of wet plastering them. The plasterboard lining to the external blockwork will match the plasterboard lining of the stud partitions and will reduce or eliminate the tendency for cracking. If the extension is built entirely from timber framing then use the same system for the internal partition walls.

- **Ceilings and roofs**: What cannot be avoided is the crack that often occurs at ceiling level where the ceiling or roof structure meets the external walls. The trick here is to cover the junction between wall and ceiling with a plaster cove to mask the join. It also helps if the size of the ceiling joists is slightly increased, so that they are not near the limit of their strength. This makes the structure more robust and reduces the deflection on the joists. If you are building a pitched roof on your extension and you want to use the roof space for storage, plan for this. Design the ceiling joists in the new roof as if they were floor joists and allow for boarding out the roof to give a proper space to put your things. This may sound obvious, but how many times have you balanced cardboard boxes full of books, suitcases, old televisions and so on on the ceiling joists in your roof space and then wondered why the ceiling below is cracking? You also run the risk of putting your foot through the ceiling when trying to find the Christmas decorations that somehow, inexplicably, have migrated to the furthest corner of the roof space!

■ **Services**: These can be an expensive part of any improvement in terms of both materials and labour. The less distance services have to travel the more economic and efficient they will be. Put similar rooms next to each other, for example bathroom next to kitchen (on the same floor) or bathroom above kitchen (on different floors). The travel distance for your hot water (from heating source to taps) can be costly over time if you have to run your taps for ages before you get hot water. This is because you have to draw off the cold water in the pipe before the hot water gets through to the taps. Then all the hot water left in the pipes will be allowed to cool, therefore wasting all that energy (and money) used to heat it up in the first place! Travel distances are also important with regard to routing of pipes and cables. Lifting of floor boards, notching of joists, 'boxing-in' of pipes or ugly pipe runs on walls can all add to the workload (and the cost) and affect the end product. Of course all these hidden services can prove problematic for any future work – how many times have you heard of a nail going through a pipe or cable? When designing floor plans try to think what furniture you will put in rooms. Consider alternative layouts because most people like to change things around now and again. So if you could put the television in, say, three different places in your new lounge, make sure that there are sufficient electrical outlets in each of the three places; do not forget that you need more than just a double socket outlet these days to cope with all the equipment (television, DVD player, satellite box, games console, floor lamp etc.), and this is especially true in children's bedrooms!

Good planning with plenty of forethought is the key to success when altering or extending your home.

Just remember : **Plan A**head

2 Legal background

INTRODUCTION

The main legal requirements which govern the control of building work (i.e. alterations, extensions and improvements to dwellings) in England and Wales are outlined in this chapter. Different control systems, regulations and Acts of Parliament often apply in Scotland and Northern Ireland and although, for example, building and planning regulations throughout the United Kingdom are all based on the same fundamental principles, there are subtle differences in form and content, and major differences in the way legislation is administered between the different countries.

Whilst there are many pieces of legislation which can apply to building work, in practice, a relatively small number of legal requirements are most frequently met, resulting in the need for

- Planning permission
- Listed building consent
- Building Regulation compliance
- Party Wall Act compliance
- Occupier's Liability Act (OLA) compliance
- Energy performance certificates (EPCs)

To further complicate matters, these different requirements are usually administered by different bodies and involve following different procedures.

With all building work, the owner of the property (or land) in question is ultimately responsible for complying with the relevant planning rules and Building Regulations (regardless of the need to apply for planning permission and/or Building Regulations approval).

Therefore, failure to comply with the relevant rules will result in the owner being liable for any remedial action (which could go as far as demolition and/or restoration). The golden rule is to always discuss the proposals with the relevant Local Planning Authority and Building Control Service before starting work.

PLANNING PERMISSION

The current planning system in England and Wales sets out conditions for development and use of land. New developments and major changes to existing buildings or the environment require planning permission. Some minor developments do not require

Extending and Improving Your Home: An Introduction, First Edition. M.J. Billington and C. Gibbs.
© 2012 M. J. Billington and C. Gibbs. Published 2012 by Blackwell Publishing Ltd.

planning permission, as they would have little or no impact on the local environment. These developments are known as permitted development. Some common developments are discussed below where the conditions and limits that might allow them to be considered as permitted development are discussed. A list summarising common activities carried out in and around dwellings is given in Table 2.1. It is very important to note that whilst many of these activities do not normally require planning consent, if carried out to a listed building then listed building consent (a form of planning permission) may still be needed.

Extensions and additions to houses (including conservatories)

An extension or addition to a house (a conservatory falls under the same planning conditions as any extension or addition to a house) is considered to be permitted development, not requiring an application for planning permission, provided that it complies with the following limits and conditions:

- No more than half the area of land around the original house may be covered by additions or other buildings. The term 'original house' means the house as it was first built or as it stood on 1 July 1948 (if it was built before that date). For example, although the current owners may not have built an extension to the house, a previous owner may have done so.
- There may be no extension forward of the principal elevation or side elevation fronting a highway.
- Any proposed extension may not be higher than the highest part of the roof.
- The maximum depth of a single storey rear extension may be no more than 3 m beyond the rear wall for an attached house and 4 m beyond the rear wall for a detached house.
- The maximum height of a single storey rear extension may be no more than 4 m.
- The maximum depth of a rear extension of more than one storey may be no more than 3 m beyond the rear wall.
- Two storey extensions may be no closer than 7 m to rear boundary.
- The roof pitch of any extensions higher than one storey should match the roof pitch of the existing house as far as is reasonably practicable.
- The maximum eaves height of an extension within 2 m of the boundary may be no more than 3 m.
- The maximum eaves and ridge height of extension may be no higher than existing house.
- Any extensions to the side of a property may only be single storey with maximum height of 4 m and width no more than half that of the original house.
- The materials used for the exterior of the extension should be similar in appearance to the existing house.
- Verandas, balconies or raised platforms are not considered to be permitted development so would need planning permission.
- Upper-floor, side-facing windows must be obscure-glazed and non-opening apart from any part of the window opening which is at least 1.7 m above the floor.

There are additional restrictions that apply if the house being extended or added to is on designated land. Designated land includes National Parks and the Broads, Areas of

Table 2.1 Planning permission and some common activities linked to the use of dwellings.

Activity	Comments
Adverts and signs	Advertisement consent may be needed to display an advertisement bigger than 0.3 m² (or any size if illuminated) on the front of, or outside, of a property (be it a house or business premises).
Basements	See detailed notes.
Biomass	Planning permission is not normally needed when installing a biomass system in a house if the work is all internal. The installation of a flue outside is normally permitted development, unless it affects the roof line, at which point planning permission should be sought.
Boilers and heating	Planning permission is not normally required for installation or replacement of a boiler or heating system if all the work is internal.
Ceilings and floors	Planning permission is not generally required to replace a floor or ceiling.
Conservatories	See extensions
Decking	Putting up decking, or other raised platforms, in a garden is permitted development, providing ■ the decking is no more than 30 cm above the ground; ■ together with other extensions, outbuildings etc., the decking or platforms cover no more than 50% of the garden area.
Decorations	Planning permission not required for internal decorations
Demolition	See detailed notes.
Doors and windows	Planning permission is not normally required for repairing, fitting or replacing doors and windows (including double glazing).
Drains and sewers	It is not usually necessary to apply for planning permission for repairs or maintenance on drainpipes, drains and sewers. Occasionally, it may be necessary to apply for planning permission for some of these works because the council has made an Article 4 Direction withdrawing permitted development rights.
Electrics	Planning permission is not generally required for installing or replacing electrical circuits.
Extensions	See detailed notes.
External walls	Planning permission is not needed for repairs, maintenance or minor improvements, such as painting a house.
Fascias	Maintenance of fascia such as replacement or painting generally does not require planning permission.
Fences, gates and garden walls	See detailed notes.
Flats and maisonettes	See detailed notes.
Flue, chimney or soil and vent pipe	See detailed notes.
Fuel tanks	See detailed notes.
Garage conversion	Planning permission is not usually required, providing the work is internal and does not involve enlarging the building. Sometimes permitted development rights have been removed from some properties with regard to garage conversions and therefore the local planning authority should be contacted before proceeding, particularly if the house is on a new housing development or in a conservation area.

Table 2.1 (*Continued*)

Activity	Comments
Gate ornaments	See detailed notes.
Heat pumps	Installing a ground source or water source heat pump system does not usually need planning permission and should fall within permitted development rights. Air-source heat pumps currently require planning permission. However, on 17 November 2009 the UK Government published a consultation on proposals for removing this requirement and extending permitted development rights to air-source heat pumps. The consultation closed on 9 February 2010 and the outcome is still awaited.
Insulation	Planning permission is not normally required for fitting insulation (where there is no change in external appearance).
Internal walls	Planning permission is not needed for internal alterations including building or removing an internal wall.
Kitchens and bathrooms	A planning application for installing a kitchen or bathroom is generally not required unless it is part of a house extension.
Lighting	Light itself, and minor domestic light fittings, are not subject to planning controls.
Loft conversion	See detailed notes.
Micro-combined heat and power	Planning permission is not normally needed when installing a micro-combined heat and power system in a house if the work is all internal. If the installation requires a flue outside see detailed notes. If the project also requires an outside building to store fuel or related equipment the same rules apply to that building as for other extensions and garden outbuildings.
Outbuildings	See detailed notes.
Patio and driveway	See detailed notes.
Paving your front garden	See detailed notes.
Porches	See detailed notes.
Roof	See detailed notes.
Satellite, TV and radio antenna	See detailed notes.
Shops	Planning permission is needed to change a residential property to a shop. This is beyond the scope of this book.
Solar panels	See detailed notes.
Trees and hedges	Many trees are protected by tree preservation orders, which means that, in general, council consent is needed to prune or fell them. In addition, there are controls over many other trees in conservation areas. The local authority should be contacted if there is any doubt about the status of trees at a property especially if it is intended to prune or fell them.
Underpinning	Maintenance on foundations generally does not require planning permission.
Wind turbines	See detailed notes.
Working from home	See detailed notes.

Important Note: in all of the above, Listed Building Consent may be needed if the work is to a listed building or the building is in a Conservation Area.

Outstanding Natural Beauty, conservation areas and World Heritage Sites. In these areas the above list is further restricted as follows:

- Rear extensions may be of no more than one storey.
- There must be no cladding of the exterior of the house or any existing extensions with stone, artificial stone, pebble dash, render, timber, plastic or tiles.
- There may be no side extensions.

Flats and maisonettes

The permitted development allowances described above apply to houses not flats, maisonettes or other buildings. For example, planning permission would be required for the single storey extension of a ground floor flat. For a loft conversion of a top floor flat, planning permission may not be required if it only involves internal works. However, local interpretation can vary and contact should be made with the local planning authority for advice.

Flue, chimney or soil and vent pipe

Fitting, altering or replacing an external flue, chimney or soil and vent pipe is normally considered to be permitted development, not requiring planning consent, if the conditions outlined below are met:

- Flues on the rear or side elevation of the building are allowed to a maximum of 1 m above the highest part of the roof.
- If the building is listed or in a designated area even it is advisable to check with the local planning authority before a flue is fitted.
- In a designated area, the flue should not be fitted on the principal or side elevation that fronts a highway.

Fences, gates, ornamental gateposts and garden walls

Planning permission is needed if it is intended to erect or add to a fence, wall or gate and

- It would be over 1 m high and next to a highway used by vehicles (or the footpath of such a highway) or over 2 m high elsewhere.
- The householder's right to put up or alter fences, walls and gates is removed by an Article 4 direction or a planning condition (always check with the local planning authority first).
- The house is a listed building or in the curtilage of a listed building.
- The fence, wall or gate, or any other boundary involved, forms a boundary with a neighbouring listed building or its curtilage.

Planning permission is not needed if it is intended to take down a fence, wall, or gate, or to alter, maintain or improve an existing fence, wall or gate (no matter how high) provided that its height is not increased. In a conservation area, however, conservation area consent might be needed to take down a fence, wall or gate.

Generally, planning permission is not needed for hedges, as such, though if a planning condition or a covenant restricts planting (e.g. on 'open plan' estates, or where a driver's sight line could be blocked) planning permission and/or other consents may be needed.

Fuel tanks

Installing a fuel tank is considered to be permitted development, not needing planning permission, subject to the following limits and conditions:

- The tank must have a capacity of not more than 3500 L.
- It must not be forward of the principal elevation fronting a highway.
- Maximum overall height permitted is 3 m; however, if it is within 2 m of a boundary this is reduced to 2.5 M.
- Not more than half the area of land around the original house (see definition above) would be covered by additions or other buildings.
- In National Parks, the Broads, Areas of Outstanding Natural Beauty and World Heritage Sites the maximum area to be covered by buildings, enclosures, containers and pools more than 20 m from a house is limited to $10\,m^2$.
- It must not be sited at the side of properties on designated land.
- Within the curtilage of listed buildings any container will require planning permission.

The permitted development regime includes liquid petroleum gas tanks as well as oil storage tanks.

Demolition

In most cases planning permission is not needed to knock down a building, unless the council has made an Article 4 direction restricting the permitted development rights that apply to demolition.

If it is decided to demolish a building, even one which has suffered fire or storm damage, it does not automatically follow that planning permission will be granted to build any replacement structure or to change the use of the site.

Where demolition of any kind of building is proposed, the council may wish to agree the details of how it is intended to carry out the demolition and how it is proposed to restore the site afterwards.

It will be necessary to apply for a formal decision on whether the council wishes to approve these details before the demolition commences. This is what is called a 'prior approval application' and the council can be contacted to explain what it involves.

Basements

The planning regime covering the creation of living space in basements is evolving and under review.

Converting an existing residential cellar or basement into a living space is, in most cases, unlikely to require planning permission, as long as it is not a separate unit or unless the usage is significantly changed or a light well is added, which alters the external appearance of the property. Excavating to create a new basement which involves major works, a new separate unit of accommodation and/or alters the external appearance of the house, such as adding a light well, is likely to require planning permission.

Loft conversions

Planning permission is not normally required. However, permission is required where the roof space is altered or extended and it exceeds specified limits and conditions.

A loft conversion for a house is considered to be permitted development, not requiring an application for planning permission, subject to the following limits and conditions:

- A volume allowance of 40 m³ additional roof space for terraced houses (remember that any previous roof space additions must be included within the volume allowances even if these have been created by previous owners).
- A volume allowance of 50 m³ additional roof space for detached and semi-detached houses.
- There must be no extension beyond the plane of the existing roof slope of the principal elevation that fronts the highway.
- The loft conversion must not be higher than the highest part of the roof.
- Materials used externally must be similar in appearance to the existing house.
- Any verandas, balconies or raised platforms are not considered to be permitted development, so would need planning permission.
- Side-facing windows must be obscure-glazed and non-opening apart from any part of the window opening which is at least 1.7 m above the floor.
- Roof extensions are not considered to be permitted development in designated areas.
- Roof extensions, apart from hip to gable ones, should be set back, as far as practicable, at least 20 cm from the original eaves.

Outbuildings

Rules governing outbuildings apply to sheds, greenhouses and garages as well as other ancillary garden buildings, such as swimming pools, ponds, sauna cabins, kennels, enclosures (including tennis courts), and many other kinds of structure for a purpose incidental to the enjoyment of the dwellinghouse.

Outbuildings are considered to be permitted development, not needing planning permission, subject to the following limits and conditions:

- There should be no outbuilding on land forward of a wall forming the principal elevation.
- Outbuildings and garages must be single storey with maximum eaves height of 2.5 m and maximum overall height of 4 m with a dual pitched roof or 3 m for any other roof.
- Outbuildings should have a maximum height of 2.5 m in the case of a building, enclosure or container within 2 m of a boundary of the curtilage of the dwellinghouse.
- Any verandas, balconies or raised platforms connected to the outbuildings are not considered to be permitted development so would need planning permission.

- No more than half the area of land around the original house may be covered by additions or other buildings.
- In National Parks, the Broads, Areas of Outstanding Natural Beauty and World Heritage Sites the maximum area to be covered by buildings, enclosures, containers and pools more than 20 m from house is limited to 10 m^2.
- On designated land buildings, enclosures, containers and pools at the side of properties will require planning permission.

Patios, paving and driveways

Specific rules apply for householders wanting to pave over their front gardens.

Planning permission is not needed if a new or replacement driveway of any size uses permeable (or porous) surfacing which allows water to drain through, such as gravel, permeable concrete block paving or porous asphalt, or if the rainwater is directed to a lawn or border to drain naturally.

If the surface to be covered is more than 5 m^2, planning permission will be needed for laying traditional, impermeable driveways that do not provide for the water to run to a permeable area.

Elsewhere around a house there are no restrictions on the area of land which can be covered with hard surfaces at, or near, ground level. However, significant works of embanking or terracing to support a hard surface might need a planning application.

Porches

The planning rules for porches are applicable to any external door to the dwellinghouse.

Adding a porch to any external door of a house is considered to be permitted development, not requiring an application for planning permission, provided the following:

- The ground floor area (measured externally) would not exceed 3 m^2.
- No part would be more than 3 m above ground level (the height needs to be measured in the same way as for a house extension).
- No part of the porch would be within 2 m of any boundary of the dwellinghouse and the highway.

Roof alterations

Planning permission is not needed to reroof a house or to insert roof lights or skylights.

The permitted development rules allow for roof alterations without the need for planning permission, subject to the following limits and conditions:

- Any alteration must project no more than 150 mm from the existing roof plane.
- No alteration must be higher than the highest part of the roof.
- Side-facing windows must be obscure-glazed and non-opening apart from any part of the window opening which is at least 1.7 m above the floor.
- **Protected species:** – work on a loft or a roof may affect bats. Protected species need to be considered when planning work of this type. A survey may be needed and, if bats are using the building, a licence may be needed. Further information is available from Natural England (www.naturalengland.org.uk).

Satellite, TV and radio antennas

Before it is intended to buy or rent an antenna, a check should be carried out as to whether planning permission, listed building consent, or permission from the landlord or owner is needed. The householder is responsible for placing antennas in the appropriate position.

The planning permission and permitted development regimes for antenna are shown below for buildings up to 15 m high. Buildings above this size are beyond the scope of this book although guidance can be found at www.planningportal.gov.uk. This link also contains a good practice guidance on installing an antenna (which contains supplementary advice on installation).

Under the Town and Country Planning (General Permitted Development) Order 1995 (as amended), there is a general permission to install antennas up to a specific size on property without the need for planning permission. This general permission depends on the house type and area. The local planning authority can give more advice.

Houses and buildings up to 15 m high

Unless the house is in a designated area, planning permission is not needed to install an antenna on a property, provided the following:

- There will be no more than two antennas on the property overall. (These may be on the front or back of the building, on the roof, attached to the chimney, or in the garden.)
- If a single antenna is being installed, it is not more than 100 cm in any linear dimension (not including any projecting feed element, reinforcing rim, mounting and brackets).
- If two antennas are being installed, one is not more than 100 cm in any linear dimension, and the other is not more than 60 cm in any linear dimension (not including any projecting feed element, reinforcing rim, mounting and brackets).
- The cubic capacity of each individual antenna is not more than 35 L.
- An antenna fitted onto a chimney stack is not more than 60 cm in any linear dimension.
- An antenna mounted on the roof only projects above the roof when there is a chimney-stack. In this case, the antenna should project more than 60 cm above the highest part of the roof or above the highest part of the chimney stack, whichever is lower.

Houses and buildings up to 15 m high in designated areas

If the house is in a designated area, planning permission is not needed to install an antenna on the property, provided that the above requirements are met and, additionally

- there will be no more than two antennas on the property overall and
- an antenna is not installed on a chimney, wall, or a roof slope which faces onto, and is visible from, a road or a Broads waterway. (Where there is doubt about the positioning, advice may be obtained from the local planning authority.)

Solar panels

In many cases fixing solar panels to a roof is likely to be considered permitted development under planning law with no need to apply for planning permission. There are, however, important exceptions and provisos which must be observed.

Leaseholders may need to get permission from the landlord, freeholder or management company.

All solar installations are subject to the following conditions:

- Panels on a building should be sited, so far as is practicable, to minimise the effect on the appearance of the building.
- They should be sited, so far as is practicable, to minimise the effect on the amenity of the area.
- When no longer needed for microgeneration they should be removed as soon as possible.

Roof and wall-mounted solar panels

The following limits apply to roof and wall-mounted solar panels:

- Panels should not be installed above the ridgeline and should project no more than 200 mm from the roof or wall surface.
- If the property is a listed building, installation is likely to require an application for listed building consent, even where planning permission is not needed.

Wall-mounted panels only

If the property is in a conservation area, or in a World Heritage Site, planning consent is required when panels are to be fitted on the principal or side elevation walls and they are visible from the highway. If panels are to be fitted to a building in the garden or grounds they should not be visible from the highway.

Stand-alone solar panels

The following limits apply to stand-alone solar panels:

- They should be no higher than 4 m.
- They should be at least 5 m from boundaries.
- The size of the array is limited to $9 \, m^2$ or 3 m wide and 3 m deep.
- They should not be installed within the boundary of a listed building.
- In the case of land in a conservation area or in a World Heritage Site they should not be visible from the highway.
- Only one stand-alone solar installation is permitted.

Wind turbines

The planning regime for installing wind turbines is complex and evolving. At present in most cases planning permission is needed from the local authority to add a domestic wind turbine to a house, or grounds surrounding the house. It is up to each local authority to decide what information must be provided with the application. It is advisable to contact the relevant local authority before applying, to discuss the following planning issues:

- Visual impact
- Noise

- Vibration
- Electrical interference (with TV aerials)
- Safety

The local planning authority should always be contacted about planning issues before a system is installed.

On 17 November 2009, the Government published proposals for extending permitted development rights to domestic wind turbines. The consultation closed on 9 February 2010 and the outcome is awaited.

Working from home

Although this subject is not directly covered by this book, it is of general interest given the number of people that now take the option of working from home and the subject often arises in discussions with clients considering alterations to their homes. The following general advice is adapted from the UK Government's Planning Portal web site.

Planning permission is not necessarily needed to work from home. The key test is whether the overall character of the dwelling will change as a result of the business.

If the answer to any of the following questions is 'yes', then permission will probably be needed:

- Will the home no longer be used mainly as a private residence?
- Will the business result in a marked rise in traffic or people calling?
- Will the business involve any activities unusual in a residential area?
- Will the business disturb neighbours at unreasonable hours or create other forms of nuisance such as noise or smells?

Whatever business is carried out from the home – whether it involves using part of it as a bed-sit or for 'bed and breakfast' accommodation, using a room as a personal office, providing a child-minding service, for hairdressing, dressmaking or music teaching, or using buildings in the garden for repairing cars or storing goods connected with a business – the key test is: is it still mainly a home or has it become business premises?

If there is any doubt an application can be made to the local council for a Certificate of Lawful Use for the proposed activity, to confirm it is not a change of use and still the lawful use.

Making an application

The easiest way to make a planning application is online via the Government's Planning Portal. The Planning Portal is the Government's official planning web site. Every local authority in England and Wales accepts planning applications via the Planning Portal.

There are many benefits to creating and submitting applications online. It can be used to complete applications for consents including

- Planning permission
- Lawful development certificates
- Listed building consent
- Conservation area consent

Completing a form online ensures that applicants are prompted to answer only questions relevant to their application. The completed form is sent online via the Planning Portal directly to the local planning authority for processing, although the applicant will need to register with the Planning Portal before using the application service. Alternatively, the standard forms can be downloaded in paper format either from the Planning Portal or the relevant local planning authority web site and then sent in by post. The relevant link is www.planningportal.gov.uk/PpApplications/genpub/en/Ecabinet. Full guidance notes to making an application are contained on the web site. Fees are charged for applications and these can be viewed online at the Planning Portal web site. Alternatively, the local planning authority should be contacted to provide fee details.

LISTED BUILDINGS AND CONSERVATION AREAS

Listed buildings

A 'listed building' is a building, object or structure that has been judged to be of national importance in terms of architectural or historic interest and included on a special register, called the List of Buildings of Special Architectural or Historic Interest. This list is compiled by the UK Department for Culture, Media and Sports (DCMS), under the provisions of the *Planning (Listed Buildings and Conservation Areas) Act 1990* (LBCA Act). The list includes a wide variety of structures, from castles, old houses and cathedrals to bridges and monuments in grave yards. It is important to note that the term 'listed building' includes

- The building itself;
- Any object or structure fixed to it and
- Any object or structure that has been within the curtilage of the building since 1948.

Why are buildings listed?

Buildings are listed to help protect the physical evidence of our past, including buildings which are valued and protected as a central part of our cultural heritage and our sense of identity. Historic buildings also add to the quality of our lives, being an important aspect of the character and appearance of our towns, villages and countryside.

What are the criteria for a building having listed status?

The Department of Culture, Media and Sport uses the following criteria to decide which buildings to include on the list of protected buildings:

- **Architectural interest:** Buildings of importance because of their design, decoration and craftsmanship.
- **Historic interest:** Buildings which illustrate an aspect of the nation's social, economic, cultural or military history.
- **Historic association:** Buildings that demonstrate close historical association with nationally important people or events.
- **Group value:** Buildings that form part of an architectural ensemble, such as squares, terraces or model villages.

In broad terms, buildings that are eligible for listed status are

- All buildings built before 1700 that survive in anything like their original condition;
- Most buildings of 1700–1840, although selection is necessary;
- Between 1840 and 1914 only buildings of definite quality and character; the selection is designed to include the major works of principal architects;
- Between 1914 and 1939 selected buildings of high quality or historic interest and
- A limited number of outstanding buildings after 1939, but at least 10 years old, and usually more than 30 years old.

Grades of listed buildings

Listed buildings are classified into grades:

- **Grade I:** Buildings of exceptional interest (approximately 2% of all listed buildings).
- **Grade II*:** Particularly important and more than special interest (approximately 4%).
- **Grade II:** Buildings of special interest, warranting every effort being made to preserve them (94%).

What part of the building is listed?

When a building is listed, it is listed in its entirety, which means that both the exterior and the interior are protected. In addition, any object or structure fixed to the building, and any object or structure within the curtilage of the building, which although not fixed to the building, forms part of the land and has done so since before 1 July 1948, are treated as part of the listed building.

Listed building consent

Listed building control is a type of planning control, which protects buildings of special architectural or historical interest. These controls are in addition to any planning regulations which would normally apply. Listed building status can also result in the requirement for planning permission where it would not ordinarily be required – for example, the erection of means of enclosure.

This special form of control is intended to prevent the unrestricted demolition, alteration or extension of a listed building without the express consent of the local planning authority or the Secretary of State. The controls apply to any works for the demolition, alteration or extension of a listed building, which are likely to affect its character as a building of special architectural or historical interest. Unless the works of demolition, alteration or extension of a listed building are authorised by means of an application to the local planning authority (Section 7 of the LBCA Act) then the person carrying out those works is guilty of an offence.

It may also be necessary to obtain listed building consent for any works to separate buildings within the grounds of a listed building. If an application for listed building consent is refused, granted with conditions or not determined within 8 weeks of it being validated by the council then there is a right to appeal to the Secretary of State.

Listed building consent may be granted subject to conditions with respect to

- Preservation of particular features of the building, either as part of it or after it is removed;
- Making good of any damage caused to the building by the works after work is completed and
- Reconstruction of the building or any parts of it following the proposed works, using the original materials as far as possible, and any alterations within the building as laid down in the conditions.

An application for listed building consent can be made after work to a listed building has taken place. However, work is only authorised from the actual date the consent is given, so anyone carrying out the demolition of a listed building or altering or extending in a way which would affect its character prior to this can still be prosecuted.

Examples of the types of alteration or extension which would normally require listed building consent are

- An extension to a building whether or not it is within the permitted development limits of the Town and Country Planning General Permitted Development Order 1995;
- Alteration such as the removal and replacement of doors and windows and
- Alterations to the interior fabric of a listed building.

The easiest way to make an application for listed building consent is online via the Government's Planning Portal. The Planning Portal is the Government's official planning web site. Every local authority in England and Wales accepts listed building consent applications via the Planning Portal. This has been discussed more fully under the section on planning permission previously.

Conservation areas

Local planning authorities hold details of conservation areas in their region. Anyone in doubt as to whether their house falls into such an area should contact their local planning authority.

If the property lies in a conservation area, then conservation area consent will be needed to do any of the following:

- Demolish a building with a volume of more than $115 \, m^3$; there are a few exceptions details of which can be obtained from the council.
- Demolish a gate, fence, wall or railing over 1 m high next to a highway (including a public footpath or bridleway) or public open space, or over 2 m high elsewhere.

Even if the proposals do not include the work mentioned above, a check should still be made to see if an application for conservation area consent is required.

If an application for conservation area consent is refused, granted with conditions or not determined within 8 weeks of it being validated by the council then you there is a right to appeal to the Secretary of State.

The easiest way to make an application for conservation area consent is online via the Government's Planning Portal. The Planning Portal is the Government's official planning web site. Every local authority in England and Wales accepts conservation area consent applications via the Planning Portal. This has been discussed more fully under the section on planning permission previously.

THE BUILDING ACT 1984 AND THE BUILDING REGULATIONS

The Building Regulations 2010 are what is known as a Statutory Instrument (sometimes called secondary legislation). Statutory Instruments contain technical information and considerable amounts of detail and are usually made under other primary legislation or Acts of Parliament. Therefore, the 2010 Regulations are made under powers contained in the Building Act 1984. The Act has been revised by other Acts of Parliament on numerous occasions and an updated version of the complete Act can be found in Knight's Guide to Building Control Law and Practice.

The Building Act covers the making and administration of Building Regulations (both the local authority and approved inspector systems) and also deals with a range of other provisions, such as drainage and the local authority's powers in relation to dangerous structures, defective premises, demolitions, the rules governing control of contravening work, the levying of fines and so on.

Both the Building Regulations and the Building (Approved Inspectors etc.) Regulations are made under powers contained in Section 1 of the Building Act and constitute the principle pieces of secondary legislation in the overall structure. The legal structure of the current building control system in England and Wales is summarised in Figure 2.1. The practical guidance mentioned in Figure 2.1 is covered in more detail in this chapter and in the practical details that form the technical chapters of this book.

BUILDING REGULATIONS

Although we may not be aware of it, the influence of the Building Regulations is around us all of the time. In our homes Building Regulations affect and control the

- Size and method of construction of foundations, walls (both internal and external), floors, roofs and chimneys;
- Size and position of stairs, room exits, corridors and external doors;
- Number, position, size and form of construction of windows and external doors (including glazing);
- Methods for disposing of solid waste;
- Design, construction and use of the services such as
  - o Above and below ground foul drainage taking the waste from kitchen and bathroom appliances (including the design and siting of the appliances themselves),
  - o Rainwater disposal systems including gutters and downpipes from roofs and drainage from paths and paving,
  - o Electrical installations,
  - o Heating and hot water installations using gas, oil or solid fuel,
  - o Fire detection and alarm systems and
  - o Mechanical ventilation systems;

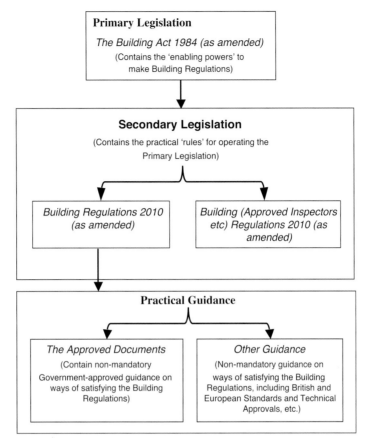

Figure 2.1 Legal structure of building control in England and Wales.

- Design and construction of the paths outside the house that
  - Lead to the main entrance and
  - Are used to access the place where refuse is stored.

Application of the Building Regulations

Put simply, the regulations apply to 'building work'. In the 2010 Building Regulations 'building work' means

(1) Erecting or extending a building;
(2) Providing or extending a controlled service or fitting;
(3) Materially altering a building, or a controlled service or fitting;
(4) Carrying out work which causes a material change of use of a building;
(5) Inserting insulating material into the cavity wall of a building;
(6) Work involving the underpinning of a building;
(7) Work required because of a change in the energy status of a building (not applicable to domestic dwellings);

(8) Work required where the thermal elements (walls, roof, floors, windows etc.) are renovated or replaced and

(9) Work required because of consequential improvements to the energy performance of the building (not applicable to buildings of less than $1000\,\mathrm{m}^2$ floor area).

This book covers work in all these categories (apart from work categories 7 and 9), so it is important to understand a little more about the terminology used. More details of this are given in the relevant chapters later. However, some minor repair work is not covered by the Building Regulations, such as

- Replacing less than 25% of the roof tiles with the same type and weight of tile;
- Replacing less than 25% of the felt to a flat roof;
- Repointing brickwork;
- Replacing floorboards and
- Renovating less than 25% of an external wall.

Additionally, although the replacement of an existing window or external door (i.e. the fixed frame as well as the moving part) needs to comply with the Regulations, works of repair to parts of the window or door such as

- Replacing broken glass,
- Replacing a sealed double glazing unit,
- Repairing or replacing ironmongery or
- Replacing rotten framing members.

are not covered by the Regulations.

All categories of 'building work' must comply with the Building Regulations. However, when carrying out work to an existing building (such as extension or alteration) it is not always necessary to bring the existing building up to the full requirements of all parts of the Building Regulations. Having said that, work to existing buildings must be carried out so that, when the work is complete, the building is not made worse in relation to compliance with the regulations than it was before the work commenced (since it is very likely that the original building may not have complied with the current Regulations and bringing it up to current standards might be expensive and impractical).

Building regulation compliance

In general, when Building Regulation compliance is needed there are two bodies that have the necessary legal powers and to which plans, particulars and details may be submitted:

(1) Local Authority Building Control Departments.
(2) Approved inspectors.

Additionally, some types of work can be carried out and 'self-certified' as complying with the Regulations by a competent person.

To carry out work without obtaining the necessary consents is a criminal offence. This can result in the unauthorised work being removed by the local authority and, if convicted, the householder and/or the builder can face a substantial fine.

Local authority building control

As the name suggests, this form of control is exercised by the relevant local authority in the area where the work will take place. Each local authority in England and Wales (Unitary, District and London Boroughs in England and County and County Borough Councils in Wales) has a Building Control section. The local authority has a general duty to see that building work complies with the Building Regulations unless it is formally under the control of an approved inspector.

Full details of each local authority (contact details, geographical area covered etc.) can be found at www.labc-services.co.uk.

Under the local authority control system there are two different procedures that can be used to ensure compliance with the Building Regulations:

(1) The Building Notice.
(2) Submission of Full Plans.

These two procedures are illustrated in Figure 2.2.

The building notice

Under this procedure, a person who intends to carry out building work or who wants to make a material change in the use of a building may give a Building Notice to a local authority in the area where the work is situated. All works of extension and alteration to domestic dwellings can be notified to the local authority via a Building Notice, except where the work would involve building over a public sewer, in which case full plans would have to be deposited.

There is no official form of building notice. However, it must be signed by, or on behalf of, the person intending to carry out the work and must contain or be accompanied by the following information:

- The name and address of the person who intends to carry out the building work.
- A statement that the notice is given in accordance with Regulation 12(2)(a).
- A description of the proposed building work or material change of use.
- Particulars of the location of the building to which the proposal relates and the use or intended use of that building.
- For work involving the erection or extension of a building, a plan to a scale of not less than 1:1250 showing

(1) Its size and position, including its relationship to adjoining boundaries;
(2) The boundaries of the site in which the building is located (usually called the curtilage of the building), and the size, position and use of every other building or proposed building within that curtilage and
(3) The width and position of any street on or within the boundaries of its curtilage.

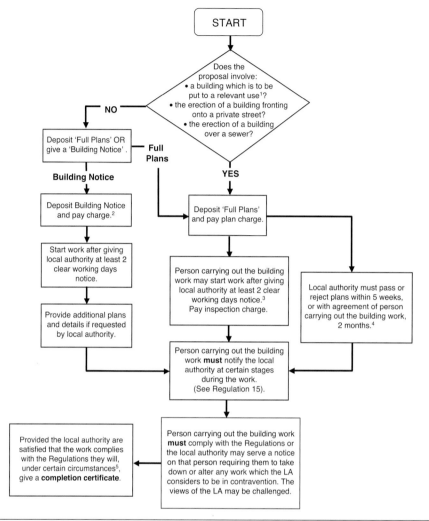

Figure 2.2   The local authority system of building control.

- A statement specifying the number of storeys (counting each basement level as one storey), in the building to which the proposal relates.
- Particulars of the provision to be made for the drainage of the building or extension and the steps to be taken to comply with any local legislation which applies.

The building notice procedure is designed to make it easier for people to apply for regulation approval for relatively minor work where detailed building plans are not really necessary (e.g. a structural alteration involving the removal of a wall to make a through lounge in a dwelling or the conversion of a spare bedroom to provide an additional bathroom). However, there may be circumstances where additional information will need to be provided so that the local authority can ensure that the regulations have been complied with. In the first example given above it may be that structural calculations are needed to confirm that the structure above the removed wall has been properly supported. Therefore, the local authority is entitled to ask for additional plans and information.

The local authority must specify in writing the information that is required and a time limit can be laid down for its provision. Curiously, although the local authority can demand the supply of additional material this is not deemed to have been 'deposited' under the regulations therefore, there are no powers to pass or reject it. Clearly, the extent of the additional information that is requested will depend on the type and complexity of the work being carried out. For example, the construction of a new dwellinghouse using the building notice procedure would undoubtedly result in a considerable amount of additional information being sought by the local authority. Since the developer, having produced this information, would not have the benefit of having his plans passed by the local authority it is probably more sensible to use the full plans procedure in such cases. As is indicated above, the building notice procedure is most useful for minor alterations and extensions to dwellings where the production of scale plans is not usually needed.

To a certain extent the popularity of the Building Notice as a form of local authority notification has diminished in recent years due to the increase in the number and scope of competent persons schemes (CPS), which has meant that many types of minor work are now carried out under these schemes and do not require a Building Notice to be deposited. Full details of the building notice procedure are given in Regulations 12 and 13 of the 2010 Regulations.

Duration

A building notice automatically ceases to have effect 3 years from the date on which it was given to the local authority if the work has not started or the change of use has not taken place. This contrasts with the situation regarding the deposit of full plans (discussed later).

Charges

Local authorities are empowered under the *Building (Local Authority Charges) Regulations 2010* to fix and to recover charges for the performance of their building regulation control functions in line with principles prescribed in the 'Charges' regulations. Usually, the charge for a building notice must be paid when the notice is given to the local authority, or the notice is not deemed to have been given. However, the local authority may agree to the charge being paid in instalments, although in this case it will lay down the amounts to be paid and the dates on which the sums will be due.

Advantages and disadvantages

From a practical viewpoint, the main advantage of using a building notice lies in the fact that work can be started as soon as the notice has been given to the local authority and the required charge has been paid, although there is still a requirement, which also applies in the case of the full plans procedure, that the local authority be given notice of at least two clear working days before work starts. The main disadvantages are that there is no procedure for approval of information supplied, the local authority is not required to acknowledge receipt of the notice (although most do) and the local authority is not obliged to issue a completion certificate when the work has been finished. Therefore, on resale of the property at a later date it may be difficult to prove to a prospective purchaser that Building Regulation compliance was sought for alterations and so on and this may delay the sale. If a notice of compliance is definitely required then the full plans route should be used.

Submission of full plans

This is the 'traditional' route for ensuring compliance with the regulations, whereby full plans and supporting information are deposited with the relevant local authority in accordance with Section 16 of the *Building Act 1984* as supplemented by Regulation 14. It is important to note that the term 'full plans' includes a great deal more than just drawings. The *Building Act 1984, Section 126* defines 'plans' as including drawings of any description and specifications or other information in any form.

Time periods and charges

Once the required information has been deposited with the local authority, the local authority has five weeks in which to pass or reject the plans. This period can be extended up to two calendar months from the date of deposit provided that the person carrying out the building work gives written consent to the extension of time. This written consent must be given before the 5 week period expires. These time periods will only commence on deposit of plans if the applicant has paid a plan charge at the same time as the plans are deposited (which may also involve the submission of a 'reasonable estimate' of the cost of the works in certain circumstances). The powers of local authorities to levy charges are contained in the *Building (Local Authority Charges) Regulations 2010*, whereby they are able to fix and recover charges for the performance of their building regulation control functions in line with principles prescribed in the 'Charges' regulations. By the same token if the local authority does not give a decision within the statutory time periods it must refund the plan charge.

Duties of a local authority to pass or reject plans

The local authority must pass the plans of any proposed work deposited in accordance with the regulations unless the plans are defective or show that the proposed work would con-travene the regulations. The term 'defective' is not defined in either the regulations or the Building Act but is usually taken to mean that insufficient or contradictory information has been supplied. If the plans cannot be passed as they stand the local authority must either

- Issue a notice of rejection detailing those parts of the regulations or Building Act which have not been complied with, or otherwise why the plans are considered to be defective or
- Pass the plans subject to conditions.

Once the plans have been passed by the local authority the work must be commenced within a period of 3 years from the date of deposit or the approval will lapse. The onus is on the local authority to give formal notice to the applicant to this effect.

Conditional passing of plans

Plans can only be passed conditionally if the applicant has issued a written request to the local authority giving consent for them to do this. The local authority may only apply the following conditions when taking this route

- That certain specified modifications be made to the deposited plans and
- That such further plans as they may specify be deposited.

Making a full plans submission

Regulation 14 specifies that plans must be deposited in duplicate, with the local authority being authorised to retain one set. It is usual for the other set to be returned to the applicant once a decision has been made although the local authority is not required by law to do this.

A full plans submission must contain the following information:

- A description of the proposed building work or material change of use, and the plans, particulars and statements required for a Building Notice.
- Details of the precautions to be taken where the work relates to the erection or extension of a building, or to works of underpinning, adjacent to or over a drain, sewer or disposal main that is shown on the sewerage records of the sewerage undertaker.
- Any other plans and information and so on that are needed to show that the work would comply with the regulations.

The additional information is often in the form of calculations, for example, to prove structural stability or thermal insulation values.

When making a full plans submission, it is possible for the person carrying out building work to request that on completion of the work they wish the local authority to issue a completion certificate. More details of this are included later.

There is no official full plans application form, although each local authority will be able to supply a potential applicant with its own personalised form. There is no legal requirement to use such a form and most local authority forms vary in detail. However, all should contain at least the information discussed above. Alternatively, a Building Regulations full plans application can be submitted online using the Local Authority Building Control (LABC) Submit-a-Plan National Portal for making electronic and offline Building Control applications to any local authority in England, Wales and Northern Ireland as well as reporting dangerous structures. The site has been designed as a single location for both the general public and professional users to submit Building Control applications directly to their intended local authority. Users can also track the progress of their application online via the site's dedicated portal DataSpace On-line. LABC can be contacted at www.submitaplan.com/ by anyone who wants to use this service.

Consultations

Once a full plans submission has been deposited with a local authority and the necessary charge has been received it will carry out a thorough check of the information provided in order to establish the extent to which the proposals comply with the regulations. In many cases this checking process will involve the local authority entering into consultation with other statutory bodies, such as the fire authority and the sewerage undertaker. For the type of work covered by this book fire authority consultation will not be needed, so only the process for the sewerage undertaker consultation is described. It should be noted that the building notice procedure cannot be used in cases where the following consultation becomes necessary.

Sewerage undertaker

Regulation 15 requires that the local authority consult the sewerage undertaker in cases where it is established that the erection or extension of a building, or works of under-pinning, will be adjacent to or over a drain, sewer or disposal main that is shown on the sewerage records of the sewerage undertaker. Consultations must take place

- As soon as practicable after the plans have been deposited and
- Before the local authority issues any completion certificate in relation to the building work.

Additionally, the local authority must

- Give the sewerage undertaker sufficient plans to show that the work would comply with the applicable requirements of paragraph H4 of Schedule 1;
- Have regard to any views expressed by the sewerage undertaker and
- Allow the sewerage undertaker 15 days in which to express its views before passing the plans or issuing a completion certificate, unless the sewerage undertaker has expressed its views to them before the expiry of that period.

The consultation procedure is necessary to ensure that the sewerage authority's legal interests (involving access to and structural stability of sewers) are preserved. Further consultation may also be necessary if the actual construction work reveals considerable variation from that shown on the plans.

Control over work in progress

Two methods of notifying the local authority have been outlined above, when there is an intention to carry out building work that is covered by the regulations. Irrespective of the method of notification used, local authorities have certain powers to inspect works in progress and are entitled to enter premises in order to enforce the Building Regulations, or any provisions of the *Building Act 1984*. To facilitate this, the onus is placed on the 'person carrying out the building work' to notify the local authority at certain stages in the construction process. The term 'person carrying out building work' is not defined in the regulations; however, decided case law would indicate that this can mean the owner of a building who authorises a contractor to carry out building works on his behalf. Additionally, the *Building Act, Section 36* enables the local authority to take action against the owner of the building in cases where the Building Regulations have been contravened. More usually, the contractor will take responsibility for notifying the local authority, since it is they who will have detailed knowledge of the construction programme. It should be noted that whilst the person carrying out the building work is contravening the regulations if they do not give the required notice, there is no equivalent legal requirement for the local authority to carry out an inspection after being notified. To counter this the *Building Control Performance Standards* set out recommendations for the level of service in respect of site inspections for all building control bodies. Local authorities that subscribe to the performance standards will have regard to this when setting out their inspection policies.

Notification methods

The stages of notification are laid down in Regulation 16. The means of giving the notices is no longer specified in the regulations; therefore it is up to the person carrying out the building work to agree with the local authority the method to be used. Most local authorities still issue pre-printed postcards with multichoice options that can be used for any of the 'statutory' notifications that are described below. This is harking back to the days when the notices had to be in writing; however, there is no reason why notification cannot be in the form of an email, text message, telephone call, fax or letter. The important thing to remember is that, whichever means of notification is used, it may be necessary at some later date to provide evidence of the date of service of the notice, especially where work has been covered up before being inspected. All the notice periods are specified in terms of days, where 'day' means any period of 24 h commencing at midnight and excludes any Saturday, Sunday, Bank Holiday or Public Holiday.

Notice of commencement

Irrespective of whether a person is using the building notice or full plans procedure, if they propose to commence controlled building work they must notify the local authority in sufficient time so that at least 2 days have elapsed since the end of the day on which they gave the notice.

This enables the local authority to check whether it has already been notified of the proposal, either by means of a building notice or a full plans submission.

If a full plans submission has been made, there is no legal necessity to wait for it to be passed before commencing work. However, any work done will be at the risk of the person carrying out the work and if, subsequently, the local authority finds that the plans are defective it would be within its rights to ask for any contravening work to be altered or removed so that it did comply. Most local authorities inspect work before the plans are passed but always state that the inspection is done without prejudice to its rights to take action if the plans are later found to be defective or show a contravention of the regulations.

Notice of works ready for inspection

Covering up of the following work must not commence until the due notice has been given to the local authority and at least one day has elapsed since the end of the day on which the notice was given

- Any excavation for a foundation;
- Any foundation;
- Any damp-proof course;

- Any concrete or other material laid over a site and
- Any drain or sewer to which the regulations apply.

The point of these notifications is to give the local authority the opportunity to inspect parts of the construction which become inaccessible once they have been covered up.

Where no notice is given or the work has been covered up before the notice period has elapsed the local authority has certain powers to require the exposure of the offending work and to have it corrected if it is not in compliance.

Notice to be given after completion of certain stages of work

In the following cases the local authority must be notified within 5 days of the work being completed

- When a drain or sewer which is subject to the requirements of Part H (Drainage and waste disposal) has been laid, haunched or covered up and
- The building work which is the subject of the building notice or full plans submission.

In the case of drains, the local authority is entitled to carry out watertightness tests (Regulation 45, *Testing of building work*).

Completion certificate

The local authority only becomes liable to issue a completion certificate when a notice is served on it under Regulation 17 stating that the building work has been completed and when the local authority is satisfied, after having taken all reasonable steps, that the relevant requirements of Schedule 1 specified in the certificate have been satisfied. As has been mentioned above, the full plans application form supplied by the local authority will usually contain a 'tick box' indicating that the applicant wishes the local authority to issue a completion certificate when the works are satisfactorily completed. This certificate will relate to all the applicable requirements of Schedule 1. Where a 'tick box' is not provided the applicant will have to formally request such a certificate. If this is not done the local authority will only be under an obligation to issue a completion certificate for work which is subject to the requirements of Part B of Schedule 1 (Fire safety). A completion certificate is a valuable document and will almost certainly be requested by the purchaser's solicitor if the building is subsequently sold.

Testing of building work and sampling of materials

Regulations 45 and 46 of the Building Regulations 2010 empower local authorities to test building work to ensure compliance with the requirements of Regulation 7 (*Materials and workmanship*) and any applicable parts of Schedule 1, and to take samples of materials to be used in the carrying out of the building work. In practice, for the type of work covered by this book the only test carried out would be to establish the airtightness/watertightness of drains and sewers. Regulation 46 allows a local authority to ask for test evidence that certain building materials, components or designs and so on will achieve the necessary standards of performance to comply with the regulations. For example, manufacturer's test data could be supplied by a designer to prove that they met the requirements for fire resistance and surface spread of flame for a material or component, and it is common to supply test data for strength of materials, such as cube tests for concrete and pull-out tests for fixings to cladding panels.

Using an approved inspector

Section 49 of the Building Act 1984 defines an 'approved inspector' as being a person approved by the Secretary of State or a body designated by him for that purpose. Part 2 of the Building (Approved Inspectors etc.) Regulations 2010 sets out the detailed arrangements and procedures for the grant and withdrawal of approval.

There are two types of approved inspector:

- Corporate bodies, such as the NHBC or HCD Building Control Ltd.
- Private individuals, not firms (referred to as non-corporate approved inspectors).

Approval may limit the description of work in relation to which the person or company concerned is an approved inspector.

Obtaining approval as an approved inspector

A private individual or corporate body wishing to carry out building control functions as an approved inspector must be registered with the Construction Industry Council (CIC) under rules laid down by the Secretary of State. To be registered as an approved inspector a number of criteria must be met. These include the holding of suitable professional qualifications, demonstration of adequate practical experience and the carrying of suitable indemnity insurance. In accordance with the responsibilities entailed by CIC's appointment as the body designated to register approved inspectors, it established the Construction Industry Council Approved Inspectors Register (CICAIR) to maintain and operate the Approved Inspector Register. CICAIR provides applicants with a route to qualification as an approved inspector and, upon qualification, full Registration facilities. The CIC web site at www.cic.org.uk contains details of all currently registered approved inspectors.

Procedures

Figure 2.3 outlines the essential stages in the control procedure that may be used by a person intending to carry out any building work to which the regulations apply as an alternative to using the local authority system described in earlier in this chapter.

Instructing an approved inspector

When instructing an approved inspector the client should be aware that a formal contractual arrangement will exist, and all arrangements between the parties will be subject to the law of England and Wales. Essentially, the approved inspector is acting as a consultant who is being commissioned by the client to carry out a Building Regulations assessment service. Therefore, it is usual for the contract to contain clauses covering

- The consultant's services and obligations.
- The client's requirements and obligations.
- Intellectual property rights and confidentiality of both parties to the contract.

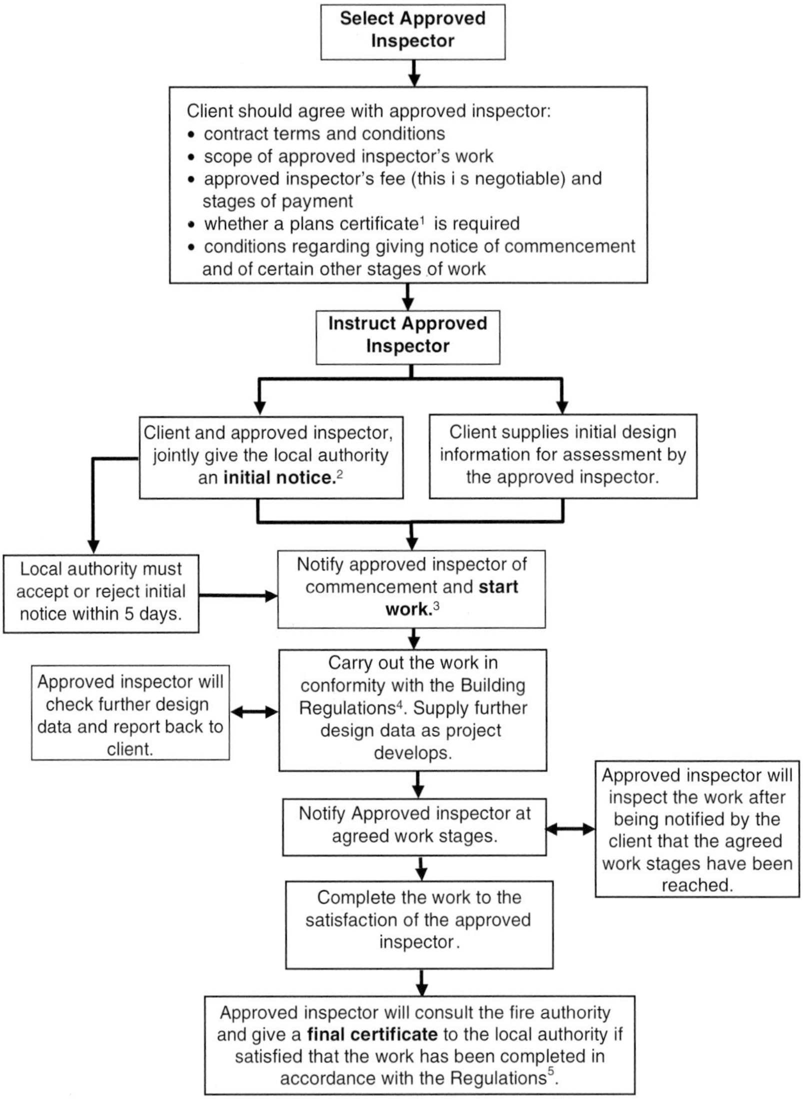
Select Approved Inspector
Client should agree with approved inspector:
• contract terms and conditions
• scope of approved inspector's work
• approved inspector's fee (this i s negotiable) and stages of payment
• whether a plans certificate1 is required
• conditions regarding giving notice of commencement and of certain other stages of work
Instruct Approved Inspector
Client and approved inspector, jointly give the local authority an initial notice.2
Client supplies initial design information for assessment by the approved inspector.
Local authority must accept or reject initial notice within 5 days.
Notify approved inspector of commencement and start work.3
Approved inspector will check further design data and report back to client.
Carry out the work in conformity with the Building Regulations4. Supply further design data as project develops.
Notify Approved inspector at agreed work stages.
Approved inspector will inspect the work after being notified by the client that the agreed work stages have been reached.
Complete the work to the satisfaction of the approved inspector.
Approved inspector will consult the fire authority and give a final certificate to the local authority if satisfied that the work has been completed in accordance with the Regulations5.

1 The person carrying out the building work may also request that the approved inspector supplies a plans certificate a copy of which must also be supplied to the local authority. Possession of a plans certificate can give valuable protection in the event that the initial notice is cancelled or ceases to be in force and no new initial notice is given or accepted. A plans certificate may be combined with the initial notice or given subsequently. If the approved inspector is unable to give a plans certificate an application may be made to the Secretary of State for a determination and a fee is payable. It may be necessary for the approved inspector to consult the fire authority before giving a plans certificate.

2 An initial notice must be accompanied by certain details and by evidence that the approved inspector is suitably insured. It may be necessary for the approved inspector to consult the fire authority and the sewerage undertaker at this stage. If the initial notice is cancelled the work must not be continued with unless it is being supervised either by a new approved inspector or by the local authority. Another approved inspector may be engaged provided that the local authority has not taken positive steps to supervise the work.

3 Work may be started once the initial notice has been expressly accepted by the local authority (or is deemed to have been accepted after 5 days have elapsed without its being rejected). Work should not be started if the notice is rejected.

4 The work must comply with the Regulations. The approved inspector is empowered to give a written notice to the person carrying out the building work if he considers that the work contravenes the Regulations. Failure to remedy the contravention within 3 months of such a notice will lead to cancellation of the initial notice by the approved inspector.

5 The local authority has 10 days in which to reject the final certificate on grounds specified within the approved inspector regulations

Figure 2.3 The approved inspector system of building control.

- Details of the consultant's insurance requirements.
- The status of the consultant regarding conflicts of interest. Except for minor work, the approved inspector is not permitted to have a professional or financial interest in the building work.
- Questions arising between the consultant and the client regarding conformity of the plans with the Building Regulations. Normally, any disputes would be referred to the Secretary of State for a determination.
- The methods for dealing with contraventions, relaxations and dispensations.
- Methods for dealing with the transmission of notices under the contract.

Level of service

The level of service offered may vary between approved inspectors, but it will always meet the standards laid down in the Building Control Performance Standards for approved inspectors who are members of the ACAI. Therefore, all such approved inspectors will check that the design complies with

- all relevant parts of the Building Regulations (including carrying out checks on structural, thermal and fire engineering design calculations, if these form part of the design details) and
- all sections of Acts of Parliament which constitute powers linked to Building Regulations under which plans can be rejected.

Additionally, there are certain statutory provisions (London Building Acts, Local Acts of Parliament etc.) which remain the responsibility of the relevant local authority to enforce. Requirements laid down by local authorities under such legislation can affect Building Regulations assessment decisions made by the approved inspector. Therefore, most approved inspectors will offer to negotiate with the relevant local authority on behalf of the client for compliance with such provisions. As this service will normally attract an additional fee, it should be referred to explicitly in the contractual agreement with the approved inspector.

Formal instruction

Once negotiations have been completed and the contract arrangements have been settled the approved inspector will require a formal instruction from the client. Most approved inspectors can provide the client with a standard Instruction Notice which when received and acknowledged by the approved inspector formally establishes the contractual relationship between the parties for each particular project (or series of projects) as appropriate.

Minor work

An approved inspector must have no professional or financial interest in the work supervised, unless it is 'minor work'. For example, this prevents an approved inspector from supervising work that he or she has designed or constructed, or from working for a company which has an interest in the work.

This exclusion does not apply to what is defined as 'minor work' which includes

- The extension or alteration of a single or two storey dwellinghouse (or its controlled services or fittings), provided that the dwellinghouse does not exceed three storeys after

the alterations (any basements may be disregarded when counting storeys). The definition of dwellinghouse does not include flats and
■ Work involving the underpinning of a building.

Therefore, for most of the building work covered by the scope of this book it would be quite in order for the client to appoint a consultant to carry out the design function, supervise the work on site and act as the building control supervisor provided that the consultant was a registered approved inspector and, in fact, there are a number of such people who offer this one-stop shop service for minor work.

Approved inspector building control procedures

The initial notice

When the client has appointed an approved inspector for a particular project, the first step in the approved inspector process will be the service of an initial notice on the local authority in whose area the work is to be carried out. This will make the local authority aware that building work in their area is being legally controlled under the Regulations, and will notify it of certain linked powers that they have under the *Building Act 1984* and any local Acts of Parliament. The local authority has 5 days in which to accept or reject the initial notice. Acceptance effectively suspends the local authority's powers to enforce the Building Regulations for the work described in the notice for as long as it remains in force, and the supervisory function passes to the approved inspector.

Thus, the initial notice is crucial to the operation of the approved inspector system, and great care must be taken that

■ It is completed correctly;
■ It is served on the correct office of the relevant local authority and
■ It is accepted by the local authority before any work starts on site.

The notice must be in a form prescribed by the *Building (Approved Inspectors etc.) Regulations 2010, Regulation 10, Form 1*. To simplify completion of the initial notice, much of the information required by law can be pre-printed. Indeed, it is usual practice for the approved inspector to supply the client with a blank notice which can then be filled out with the information specific to the client and the proposed building work. The partially completed notice is then returned to the approved inspector for completion and forwarding to the local authority.

Once the approved inspector has received the notice and the information referred to above from the client, they will be able to complete the notice and send the whole package to the relevant local authority. The local authority must accept or reject the notice within 5 days of receipt (if it does not do this the notice becomes accepted by default). Therefore, the approved inspector will send the notice and accompanying documents by recorded delivery so that proof of receipt can be verified. The local authority can reject the initial notice only if it is defective in terms of accuracy and completeness.

Cancellation of the initial notice

Generally, the initial notice remains in force during the currency of the works. However, in certain circumstances, it may be cancelled, or cease to have effect after the lapse of certain defined periods where there has been a failure to give a final certificate to the local authority.

In the following cases, the approved inspector must cancel the initial notice by issuing to the local authority a cancellation notice in a prescribed form:

- The approved inspector has become or expects to become unable to carry out (or continue to carry out) his/her functions.
- The approved inspector believes that because of the way in which the work is being carried out he/she cannot adequately perform his functions.
- The approved inspector is of the opinion that the requirements of the Regulations are being contravened and despite giving notice of contravention to the person carrying out the work that person has not complied with the notice within the 3-month period allowed.

It is also possible for the person carrying out the work to cancel the initial notice. This arises if it becomes apparent that the approved inspector is no longer willing or able to carry out his functions (through bankruptcy, death, illness etc.). This must be done in the prescribed form and must be served on the local authority and (where practicable) on the approved inspector.

Once the initial notice has ceased to have effect, the approved inspector will be unable to give a final certificate and the local authority's powers to enforce the Building Regulations can revive. If the local authority becomes responsible for enforcing the Regulations it must be provided, on request, with plans of the building work so far carried out. Additionally, it may require the person carrying out the work to cut into, lay open or pull down work so that it may ascertain whether any work not covered by a final certificate contravenes the Regulations. If it is intended to continue with partially completed work, the local authority must be given sufficient plans to show that the work can be completed without contravention of the Building Regulations. A fee, which is appropriate to that work, will be payable to the local authority.

Finally, a local authority may cancel an initial notice if it appears to them that the work to which the initial notice relates has not been commenced within a period of 3 years dating from when it was accepted by the local authority.

Ensuring compliance with the Building Regulations

When the initial notice has been accepted, the approved inspector is required by Regulation 8 of the Building (Approved Inspectors etc.) Regulations 2010 to '*take such steps as are reasonable to enable him to be satisfied within the limits of professional skill and care that*'

(1) Relevant building work complies with the Building Regulations.
(2) Where building work involves the insertion of insulating material into the cavity in a wall after that wall has been built, the approved inspector is not required to supervise the insertion of the material, but must state in the final certificate whether or not the material has been inserted.

In theory, each approved inspector can make whatever procedural arrangements they consider necessary to meet these requirements. In practice the *Building Control Performance Standards* make recommendations covering the provision of a plan appraisal service and require the adoption of an appropriate site inspection regime, and the Regulations require that the approved inspector consults the fire authority or sewerage undertaker at certain stages during the execution of the works, although this is rarely needed for the scope of work covered by this book.

Certain obligations regarding the supply of information are also placed on the person carrying out the building work. For example, where an extension is added to an existing dwelling, the client (i.e. the person carrying out the building work) may need to provide certain energy calculations showing that the extension will not make the energy use of the extended building any worse than it was before the extension was built.

The contract between the client and the approved inspector will lay down the ground rules for design assessments, site inspections and consultations (if needed), but the working arrangements will be agreed on a job by job basis with the client's designer and his contractor in order to fit in with the design and construction programmes.

To ensure these assessments, inspections and consultations are properly recorded the approved inspector will always follow up each event with a written report to the client, his designer or his contractor. These reports will indicate areas of non-compliance, pass on the views of consultees (if any) and explain the remedies available should there be a dispute over compliance.

Design assessment and approval of plans

The design assessment service provided by the approved inspector is covered by the *Building Control Performance Standards*, which recommend that clear information should be communicated by the approved inspector to the client regarding

- Non-compliance with the Building Regulations;
- Views of statutory undertakers;
- Any conditions (such as the provision of information by specific dates) pertaining to the informal approval of plans and
- Remedies available in the event of a dispute over compliance (such as applying to the Department for Communities and Local Government for a determination).

Most approved inspectors encourage the client to provide design information on a continuous basis and they will furnish the client with regular reports on the progress of the design assessment. There are no statutory 'approval' deadlines to be met and the approved inspector will fit in with the normal design programme, thus providing a continuous checking service.

Most clients like to be reassured that their designs do, in fact, comply with the relevant regulations. This can be done informally by the provision of a letter to that effect from the approved inspector, once the design has reached a stage from which it is unlikely to be substantially altered.

If a formal reassurance is required, the approved inspector can, at the request of the client, issue a **plans certificate** to the local authority. In order to issue a plans certificate, the approved inspector will need to certify that

- The client's plans have been checked;
- The approved inspector is satisfied that they comply with the Building Regulations and
- Any prescribed consultations have been carried out.

The certificate must be endorsed with the reference numbers of the inspected plans (although copies of the actual plans do not have to be sent in with the certificate). Other than the provision of the plans in the first place, the client has no input into the preparation and transmission of the plans certificate.

From the client's viewpoint, the advantage of possessing a plans certificate lies in the fact that if at a later stage the initial notice ceases to be effective, the local authority cannot take enforcement action in respect of any work described in the plans certificate if it has been done in accordance with those plans.

A plans certificate, when issued by an approved inspector, certifies that the design has been checked and that the plans comply with the Building Regulations. Its issue is entirely at the option of the person carrying out the work, and copies of it are sent by the approved inspector to that person and to the local authority.

The local authority has 5 working days in which accept or reject the plans certificate, but it may only reject it on certain specified grounds and, since it is not given copies of the plans on which it based, the grounds for rejection do not include non-compliance with the Building Regulations. Plans certificates may be rescinded by a local authority if the work has not started within 3 years of the acceptance date of the certificate.

Site inspections

Although there are no 'statutory inspections' in the approved inspector control system, the inspector is legally bound to ensure that the Building Regulations are complied with. Additionally, the approved inspector is contractually bound to his client and the conditions of contract will usually include the need for certain inspections to be carried out of the work in progress. Failure to carry out such inspections will amount to breach of contract, with the usual remedies being available for such breaches. Also, approved inspectors who have embraced the *Building Control Performance Standards* are required to adopt an appropriate site inspection regime.

Details of non-compliant work must be communicated promptly and clearly to the responsible person, including an indication of any measures believed to be necessary for rectification of the situation. The mechanisms for appealing against or disputing the decision of an approved inspector should also clearly be made known to the responsible person.

During the inspection phase, the approved inspector must ensure that all statutory consultees (where applicable) are notified of any significant departure from the plans.

Non-conforming work and enforcement actions

Approved inspectors are not empowered to enforce the Building Regulations (this can only be done by local authorities). However, there are powers in Part 3 of the Building (Approved Inspectors etc.) Regulations 2010 that can result in the cancellation of the initial notice and reversion of the work to the local authority, resulting in the resurrection of their full enforcement powers.

Reversion of the work to the local authority for enforcement is a final option rarely, if ever, used in practice. It will only arise where the approved inspector has tried all other courses of action without success, and feels that the extent of the contravention is so serious that the threat of cancellation of the initial notice is the only option left. In such circumstances, the approved inspector may give a notice of contravention to the client specifying

■ Those requirements of the Building Regulations which he feels are not being complied with and
■ The location of the non-compliant work.

The notice will inform the client that unless the contravention is rectified within a period of 3 months from its date of service, then the initial notice will be cancelled. This gives the client a period of grace in which to instruct the builder to take down and remove the contravening work or to carry out alterations to make it comply.

Where the contravening works have not been rectified within the 3-month period the approved inspector will cancel the initial notice. This is done by serving a notice of cancellation on the local authority and the client.

The final certificate

When the approved inspector is satisfied that the project has been completed, they must give a **final certificate** to the local authority. The local authority has 10 days in which to accept or reject the certificate, but it can only be rejected on the following prescribed grounds:

■ The certificate is not in the prescribed form.
■ It does not describe the work to which it relates.
■ No initial notice relating to the work is in force.
■ The certificate is not signed by the approved inspector who gave the initial notice or that he is no longer an approved inspector.
■ Evidence of approved insurance is not supplied.
■ There is no declaration of independence (except for minor work).

Acceptance of the final certificate effectively removes the local authority's powers to take proceedings against the client for a contravention of Building Regulations in relation to the work referred to in the final certificate. The approved inspector does not have to copy the final certificate to any other party, but it is usual for a certificate of completion to be issued to the client once the local authority has accepted the final certificate. If the local authority rejects a final certificate, the initial notice to which it refers will cease to be in force within 4 weeks of the date of rejection and, unless the reasons for rejection can be successfully addressed, the local authority's powers of enforcement will be revived.

COMPETENT PERSONS SCHEMES

Competent persons schemes were introduced by the UK Government to allow individuals and enterprises to self-certify that their work complies with the Building Regulations as an alternative to submitting a building notice or full plans under the local authority control system or using an approved inspector.

The principles of self-certification are based on giving people who are competent in their field the ability to self-certify that their work complies with the Building Regulations without the need to submit a building notice or deposit full plans and thus incur local authority or approved inspector inspections or fees.

By establishing such schemes the Government hoped that the move towards self-certification would

- Significantly enhance compliance with the requirements of the Building Regulations;
- Reduce costs for firms joining recognised schemes and
- Promote training and competence within the industry.

It was hoped that it would also help tackle the problem of 'cowboy builders', and assist local authorities with enforcement of the Building Regulations.

Benefits of competent persons schemes

The rationale behind the schemes is to authorise, on the basis of risk to health and safety, schemes whose members are adjudged sufficiently competent in their work to self-certify that their work has been carried out in compliance with all applicable requirements of the Building Regulations.

The assumed benefits offered by the schemes to consumers are that they result in

- Lower prices, as building control fees are not payable;
- Reduced delays, as the full local authority administrative procedures do not need to be followed and
- The ability to identify competent firms.

The assumed benefits offered by the schemes to firms are

- The time and expense of submitting a building notice or full plans are avoided by firms who join these schemes and
- The schemes also allow local authority building control departments to concentrate their resources on the areas of highest risk.

Joining a competent persons scheme

Apart from the 'Gas Safe Register', membership of a competent persons scheme is not compulsory. Businesses carrying out work covered by the Building Regulations may choose to join the schemes if they judge membership to be beneficial. Alternatively, they may choose to continue to use local authority Building Control or to employ a private sector approved inspector.

If a company or individual chooses to join a competent persons scheme, they are first vetted to ensure they meet the conditions of membership, including appropriate and relevant levels of competence. Each scheme has its own procedures and requirements for vetting potential members. If they meet the conditions imposed by the relevant scheme they are classified as 'competent persons'.

The work of organisations or individuals accepted as members of a scheme is not subject to Building Control inspection. Instead, the competent person self-certifies that the work is in compliance with the Building Regulations. They issue a certificate to the consumer to this effect. In some schemes they then report the work to the scheme organisers, who in turn inform the local authority that work has taken place. In yet other schemes the organisers will inform the local authority that the work has been completed and also send a Building Control Compliance Certificate to the consumer on behalf of the competent person.

Legislative background

The powers used to set up schemes are in the *Building Act 1984* as augmented by Regulations 12, 15 and 20 of the Building Regulations 2010. Regulation 12 Subparagraph (6) states that

A person intending to carry out building work is not required to give a building notice or deposit full plans where the work consists only of work

(a) *described in column 1 of the Table in Schedule 3 if the work is to be carried out by a person described in the corresponding entry in column 2 of that Table or*
(b) *described in Schedule 4.*

Schedules 3 and 4 are shown in Boxes 2.1 and 2.2. It should be noted that although the work described in Schedule 4 is not notifiable to the local authority it must still comply with the Building Regulations and the local authority retains the powers to inspect the work if it suspects that the work has been done without due regard to public health and safety or the conservation of fuel and power.

Regulation 20 places a duty on the *'person carrying out the work'* (i.e. the competent person) to notify the local authority within 30 working days of completion of the work. The notification is in the form of a certificate, which is given to the occupier of the premises, that confirms that Regulations 4 and 7 of the Building Regulations 2010 have been complied with. The competent person can choose whether to give the local authority the certificate or merely to notify them that the work is complete. It should be noted that although the local authority is authorised to accept such a certificate it is not legally bound to, since circumstances might exist whereby the local authority doubted the authenticity or veracity of the certificate or it may have reason to believe that the work did not comply with the regulations.

Where an approved inspector is engaged to carry out the building control work the notices must be given to the approved inspector in the same way as for the local authority.

In practice, the different competent persons schemes discharge this notification duty in different ways on behalf of their members. Most schemes receive notifications of completion from their members and they then notify the relevant local authorities or approved inspectors electronically in an approved form.

THE PARTY WALL ACT

The Party Wall Act 1996 came into force in 1997 and covers three different types of work:

- Alterations to a shared (party) wall.
- The construction of new walls on the boundary.
- Excavation work close to neighbouring properties.

Box 2.1 Schedule 3: Self-certification schemes and exemptions from requirement to give building notice or deposit full plans

Column 1 *Type of work*	Column 2 *Person carrying out work*
1. Installation of a heat-producing gas appliance.	A person, or an employee of a person, who is a member of a class of persons approved in accordance with regulation 3 of the Gas Safety (Installation and Use) Regulations 1998.
2. Installation of heating or hot water system connected to a heat-producing gas appliance, or associated controls.	A person registered by Ascertiva Group Limited, Association of Plumbing and Heating Contractors (Certification) Limited, Benchmark Certification Limited, Building Engineering Services Competence Accreditation Limited, Capita Gas Registration and Ancillary Services Limited, ECA Certification Limited, HETAS Limited, NAPIT Registration Limited, Oil Firing Technical Association Limited or Stroma Certification Limited in respect of that type of work.
3. Installation of (a) an oil-fired combustion appliance or (b) oil storage tanks and the pipes connecting them to combustion appliances.	A person registered by Ascertiva Group Limited, Association of Plumbing and Heating Contractors (Certification) Limited, Benchmark Certification Limited, Building Engineering Services Competence Accreditation Limited, ECA Certification Limited, HETAS Limited, NAPIT Registration Limited or Oil Firing Technical Association Limited in respect of that type of work.
4. Installation of a solid fuel burning combustion appliance.	A person registered by Ascertiva Group Limited, Association of Plumbing and Heating Contractors (Certification) Limited, Benchmark Certification Limited, Building Engineering Services Competence Accreditation Limited, ECA Certification Limited, HETAS Limited or NAPIT Registration Limited in respect of that type of work.

(Continued)

5. Installation of a heating or hot water system connected to an oil-fired combustion appliance or its associated controls.	A person registered by Ascertiva Group Limited, Association of Plumbing and Heating Contractors (Certification) Limited, Benchmark Certification Limited, Building Engineering Services Competence Accreditation Limited, ECA Certification Limited, HETAS Limited, NAPIT Registration Limited, Oil Firing Technical Association Limited or Stroma Certification Limited in respect of that type of work.
6. Installation of a heating or hot water system connected to a solid fuel burning combustion appliance or its associated controls.	A person registered by Ascertiva Group Limited, Association of Plumbing and Heating Contractors (Certification) Limited, Benchmark Certification Limited, Building Engineering Services Competence Accreditation Limited, ECA Certification Limited, HETAS Limited, NAPIT Registration Limited, Oil Firing Technical Association Limited or Stroma Certification Limited in respect of that type of work.
7. Installation of a heating or hot water system connected to an electric heat source or its associated controls.	A person registered by Ascertiva Group Limited, Benchmark Certification Limited, Building Engineering Services Competence Accreditation Limited, ECA Certification Limited, HETAS Limited, NAPIT Registration Limited, Oil Firing Technical Association Limited or Stroma Certification Limited in respect of that type of work.
8. Installation of a mechanical ventilation or air conditioning system or associated controls, which does not involve work on a system shared with parts of the building occupied separately, in a building other than a dwelling.	A person registered by Ascertiva Group Limited, Benchmark Certification Limited or Building Engineering Services Competence Accreditation Limited in respect of that type of work.
9. Installation of an air conditioning or ventilation system in a dwelling, which does not involve work on systems shared with other dwellings.	A person registered by Ascertiva Group Limited, Benchmark Certification Limited, Building Engineering Services Competence Accreditation Limited or NAPIT Registration Limited in respect of that type of work.

10. Installation of a lighting system or electric heating system, or associated electrical controls.	A person registered by Ascertiva Group Limited, Building Engineering Services Competence Accreditation Limited, ECA Certification Limited, NAPIT Registration Limited or Stroma Certification Limited in respect of that type of work.
11. Installation of fixed low or extra-low voltage electrical installations.	A person registered by Ascertiva Group Limited, Benchmark Certification Limited, British Standards Institution, Building Engineering Services Competence Accreditation Limited, ECA Certification Limited or NAPIT Registration Limited in respect of that type of work.
12. Installation of fixed low or extra-low voltage electrical installations as a necessary adjunct to or arising out of other work being carried out by the registered person.	A person registered by Ascertiva Group Limited, Association of Plumbing and Heating Contractors (Certification) Limited, Benchmark Certification Limited, Building Engineering Services Competence Accreditation Limited, ECA Certification Limited, NAPIT Registration Limited or Oil Firing Technical Association Limited in respect of that type of electrical work.
13. Installation, as a replacement, of a window, roof light, roof window or door in an existing dwelling.	A person registered under the Fenestration Self-Assessment Scheme by Fensa Ltd, or a person registered by BM Trada Certification Limited, the British Standards Institution, CERTASS Limited or Network VEKA Limited in respect of that type of work.
13A. Installation, as a replacement, of a window, rooflight, roof window or door in an existing building other than a dwelling	BM Trada Certification Limited, FENSA Limited. (This entry was added by virtue of the Building Regulations (Amendment) Regulations 2011.)
14. Installation of a sanitary convenience, sink, washbasin, bidet, fixed bath, shower or bathroom in a dwelling, which does not involve work on shared or underground drainage.	A person registered by Ascertiva Group Limited, Association of Plumbing and Heating Contractors (Certification) Limited, Benchmark Certification Limited, Building Engineering Services Competence Accreditation Limited or NAPIT Registration Limited in respect of that type of work.

(Continued)

15. Installation of a wholesome cold water supply or a softened wholesome cold water supply.	A person registered by Ascertiva Group Limited, Association of Plumbing and Heating Contractors (Certification) Limited, Benchmark Certification Limited, Building Engineering Services Competence Accreditation Limited or NAPIT Registration Limited in respect of that type of work.
16. Installation of a supply of non-wholesome water to a sanitary convenience fitted with a flushing device which does not involve work on shared or underground drainage.	A person registered by Ascertiva Group Limited, Association of Plumbing and Heating Contractors (Certification) Limited, Benchmark Certification Limited, Building Engineering Services Competence Accreditation Limited or NAPIT Registration Limited in respect of that type of work.
17. Installation in a building of a system to produce electricity, heat or cooling (a) by microgeneration or (b) from renewable sources (as defined in European Parliament and Council Directive 2009/28/EC of 23 April 2009 on the promotion of the use of energy from renewable sources).	A person registered by Ascertiva Group Limited, Association of Plumbing and Heating Contractors (Certification) Limited, Benchmark Certification Limited, British Standards Institution, Building Engineering Services Competence Accreditation Limited, ECA Certification Limited, HETAS Limited, NAPIT Registration Limited, Oil Firing Technical Association Limited or Stroma Certification Limited in respect of that type of work.
18. Insertion of insulating material into the cavity walls of an existing building.	A person registered under the Cavity Wall Insulation Self Certification Scheme by Cavity Insulation Guarantee Agency Limited in respect of that type of work.
19. Installation, as a replacement, of the covering of a pitched or flat roof and work carried out by the registered person as a necessary adjunct to that installation. This paragraph does not apply to the installation of solar panels.	A person registered by National Federation of Roofing Contractors Limited in respect of that type of work.
20. Any building work which is necessary to ensure that any appliance, service or fitting which is installed and which is described in the preceding entries in column 1 above, complies with the applicable requirements contained in Schedule 1. This paragraph does not apply to the provision of a masonry chimney.	The person who installs the appliance, service or fitting to which the building work relates and who is described in the corresponding entry in column 2 above.

Box 2.2 Schedule 4: Descriptions of work where no building notice or deposit of full plans required

1. Work consisting of
   (a) replacing any fixed electrical equipment which does not include the provision of
       (i) any new fixed cabling or
       (ii) a consumer unit;
   (b) replacing a damaged cable for a single circuit only;
   (c) re-fixing or replacing enclosures of existing installation components, where the circuit protective measures are unaffected;
   (d) providing mechanical protection to an existing fixed installation, where the circuit protective measures and current carrying capacity of conductors are unaffected by the increased thermal insulation;
   (e) installing or upgrading main or supplementary equipotential bonding;
   (f) in relation to an existing fixed building service, which is not a fixed internal or external lighting system:
       (i) replacing any part which is not a combustion appliance,
       (ii) adding an output device or
       (iii) adding a control device,
       where testing and adjustment of the work is not possible or would not affect the use by the fixed building service of no more fuel and power than is reasonable in the circumstances;
   (g) providing a self-contained fixed building service, which is not a fixed internal or external lighting system, where
       (i) it is not a combustion appliance,
       (ii) any electrical work associated with its provision is exempt from the requirement to give a building notice or to deposit full plans by virtue of Regulation 9 or 12(6)(b),
       (iii) testing and adjustment is not possible or would not affect its energy efficiency and
       (iv) in the case of a mechanical ventilation appliance, the appliance is not installed in a room containing an open-flued combustion appliance whose combustion products are discharged through a natural draught flue;
   (h) replacing an external door (where the door together with its frame has not more than 50% of its internal face area glazed);
   (i) in existing buildings other than dwellings, providing fixed internal lighting where no more than $100 \, \text{m}^2$ of the floor area of the building is to be served by the lighting;
   (j) replacing
       (i) a sanitary convenience with one that uses no more water than the one it replaces,
       (ii) a washbasin, sink or bidet,
       (iii) a fixed bath,
       (iv) a shower,
       (v) a rainwater gutter or
       (vi) a rainwater downpipe,

(Continued)

where the work does not include any work to underground drainage, and includes no work to the hot or cold water system or above ground drainage, which may prejudice the health or safety of any person on completion of the work;

(k) in relation to an existing cold water supply
    (i) replacing any part,
    (ii) adding an output device or
    (iii) adding a control device;

(l) providing a hot water storage system that has a storage vessel with a capacity not exceeding 15 L, where any electrical work associated with its provision is exempt from the requirement to give a building notice or to deposit full plans by virtue of Regulation 9 or 12(6)(b) and

(m) installation of thermal insulation in a roof space or loft space where
    (i) the work consists solely of the installation of such insulation and
    (ii) the work is not carried out in order to comply with any requirement of these Regulations.

2. Work which
    (a) is not in a kitchen, or a special location;
    (b) does not involve work on a special installation and
    (c) consists of
        (i) adding light fittings and switches to an existing circuit or
        (ii) adding socket outlets and fused spurs to an existing ring or radial circuit.

3. Work on
    (a) telephone wiring or extra-low voltage wiring for the purposes of communications, information technology, signalling, control and similar purposes, where the wiring is not a special location;
    (b) equipment associated with the wiring referred to in sub-paragraph (a) and
    (c) prefabricated equipment sets and associated flexible leads with integral plug and socket connections.

4. For the purposes of this Schedule
"kitchen" means a room or part of a room which contains a sink and food preparation facilities;
"self-contained" in relation to a fixed building service means consisting of a single appliance and any associated controls which is neither connected to, nor forms part of, any other fixed building service;
"special installation" means an electric floor or ceiling heating system, an outdoor lighting or electric power installation, an electricity generator or an extra-low voltage lighting system which is not a pre-assembled lighting set bearing the CE marking referred to in Regulation 9 of the Electrical Equipment (Safety) Regulations 1994 and
"special location" means a location within the limits of the relevant zones specified for a bath, a shower, a swimming or paddling pool or a hot air sauna in the Wiring Regulations, seventeenth edition, published by the Institution of Electrical Engineers and the British Standards Institution as BS 7671: 2008.

The Act applies independently of any requirements for, for example, planning permission, listed building consent, conservation area consent or building regulation approval for any work undertaken. Also, having these other approvals does not remove the need to comply with the requirements of the Party Wall Act.

The general principle of the Party Wall Act is that all work which might have an effect upon the structural strength or support function of the party wall or might cause damage to the neighbouring side of the wall must be notified. If there is any doubt as to whether or not compliance with the Act is necessary, advice should be sought from the relevant local authority building control office or professional surveyor/architect.

Application of the act

The Party Wall Act applies if someone is planning to do work on a relevant structure. For the purposes of the Act 'party wall' does not just mean the wall between two adjoining properties, it also covers the following:

- A wall forming part of one building only but which is on the boundary between two (or more) properties.
- A wall which is common to two (or more) properties, including the case where a wall has been built and a neighbour subsequently builds something adjoining it.
- A garden wall, where the wall is on or adjoining the boundary line and is used to separate the properties but is not part of any particular building.
- Floors and ceilings of flats and so on.
- Excavation near to a neighbouring property.

Exemptions under the act

Some work which is carried out to a party wall is exempt from compliance with the Act, for example

- Putting up shelves and wall units;
- Replastering and
- Electrical rewiring.

Relevant work

Work covered by the Act includes the following:

- The demolition and/or rebuilding a party wall.
- Increasing the height or thickness of a party wall.
- The insertion of a damp-proof course (either chemical injection or a physical damp proof course).
- Cutting into the party wall to provide support for load bearing beams.
- Underpinning a party wall.
- Excavations within 3 m of a neighbouring building where the excavation will go below the bottom of the foundations of the neighbouring building.

- Excavations within 6 m of a neighbouring building where the excavation will go below a line drawn 45° downwards from the bottom of the foundations of the neighbouring building.

The formal procedure which applies where work falls within the scope of the Act is for a Building Owner to serve notice on the Adjoining Owner detailing the works and to obtain the affected Adjoining Owner's consent. If that consent is not forthcoming the parties are deemed to be 'in dispute' under the Act and surveyors must be appointed so that the dispute can be resolved by way of an Award.

As with all work affecting neighbours, it is always better to reach a friendly agreement rather than resort to the full force of the law. Even where the work requires a notice to be served, it is better to discuss informally the intended work, consider the neighbours comments, and amend the plans (if needed) before serving the notice. All too often an Adjoining Owner only becomes aware of a neighbour's plans to extend or carry out other relevant works when a Party Wall Notice is received.

The process of serving a notice under the Party Wall Act is as follows:

- The person intending to carry out the work must serve a written notice on the owners of the adjoining property at least 2 months before the intended start of the work to every neighbouring party giving details of the work to be carried out.
- Each neighbouring party should respond in writing giving consent or registering dissent – if a neighbouring party does nothing within 14 days of receiving the notice, the effect is to put the notice into dispute.
- No work may commence until all neighbouring parties have agreed in writing to the notice (or a revised notice).

If any of the information is missing from a served noticed, it will be invalid in which case, any subsequent Award will also be invalid.

Notice requirements

The notice must include the following:

- Details of the owner(s) of the property undertaking the work.
- The address of the property.
- The names of all the owners of the adjoining property.
- A brief description of the proposed work.
- The proposed start date for the work.
- A clear statement that the notice is being served under the Party Wall etc. Act 1996.
- The date the notice is being served.
- If the notice is for excavation work, a drawing showing the position and depth of the excavation.

New boundary walls

If the planned work is a new boundary wall up to or on the boundary line, the process is similar to the above but the notice needs to be served at least one month before the planned start date of the work. Neighbouring parties must give written agreement within 14 days for walls on the boundary (or a dispute is deemed to have occurred). However, no formal agreement is needed for a wall which adjoins but does not cross the boundary line, the neighbour just needs not to object in writing.

Excavations

If the planned work is an excavation within the distance/depth covered by the Party Wall Act, the notice needs to be served at least one month before the planned start day of the work. Neighbouring parties must give written agreement within 14 days or a dispute is deemed to have occurred.

Dispute procedures

If agreement cannot be reached between neighbouring parties, the following process applies:

(1) A Surveyor or Surveyors is/are appointed to determine a fair and impartial Award, either

- an 'Agreed Surveyor' (someone acceptable to all parties) or
- each party appoints their own Surveyor to represent themselves.

   The first option should be cheaper as the costs should be reduced – the Surveyor (or Surveyors) will decide who pays the fees – usually it will be the party undertaking the work; the exception being where the owner of the adjoining property calls on the Surveyor unnecessarily. It should be noted that any Surveyor(s) must act within their statutory responsibilities and propose a fair and impartial Award.

(2) The Agreed Surveyor, or the individual Surveyors jointly, will produce an Award which must be fair and impartial to all parties.

(3) Once an Award has been made, all parties have 14 days to appeal to a County Court against the Award.

Agreement

Once there is an agreement, all work must comply with the notice. All the agreements should be retained to ensure that a record of the granted permission is kept. A subsequent purchaser of the property may wish to establish that the work was carried out in accordance with the Party Wall Act requirements.

A more comprehensive explanatory booklet including example letters for notices and responses can be downloaded from www.communities.gov.uk/publications/planningandbuilding/partywall.

ENERGY PERFORMANCE

The European Energy Performance of Buildings Directive requires an EPC to be produced and made available whenever a building is constructed, sold or made available for rent.

The long-term proposal for this legislation is to improve the energy efficiency of buildings in the United Kingdom and address issues such as sustainability, carbon emissions and climate change targets. This will also assist the Government to achieve its targets under the Kyoto Protocol.

A Domestic Energy Assessor (DEA) will provide an EPC for your home following a visit to the property and using appropriate approved software. Items included in the data collection at the property are as follows:

- Property type and age.
- Construction type and size.
- Levels of insulation.
- Heating and controls.
- Type and size of glazing.
- Any renewables (e.g. solar heating).

The purpose of an EPC is to provide an energy rating for a property that will advise any person, with an interest in that property, of its performance in relation to energy costs and impact on the environment.

EPCs contain

- Information on your home's energy use and carbon dioxide emissions and
- A recommendation report with suggestions to reduce energy use and carbon dioxide emissions.

Energy use and carbon dioxide emissions

EPCs carry ratings that compare the current energy efficiency and carbon dioxide emissions with potential figures that your home could achieve. Potential figures are calculated by estimating what the energy efficiency and carbon dioxide emissions would be if energy saving measures were put in place.

The rating measures the energy and carbon emission efficiency of your home using a grade from 'A' to 'G'. An 'A' rating is the most efficient, while 'G' is the least efficient. This is the same system that is used to rate the energy efficiency of white goods such as washing machines and fridges. The average efficiency grade to date is 'D'. All homes are measured using the same calculations, so the energy efficiency of different properties can be compared.

The recommendation report

EPCs also provide a detailed recommendation report showing what could be done to help reduce the amount of energy used and carbon dioxide emissions. The report lists

- Suggested improvements, for example fitting loft insulation;
- Possible cost savings per year, if the improvements are made and
- How the recommendations would change the energy and carbon emission rating of the property.

The recommendations in the recommendation report do not have to be acted upon. However, if it is decided to do so, it could make the property more attractive for sale or rent by making it more energy efficient. Additionally, there may be grants available to carry out improvements that save energy, like fitting loft insulation. These grants are discussed in more detail in Chapter 13.

OCCUPIERS' LIABILITY ACT

The Occupiers' Liability Act is governed by two Acts of Parliament, namely

- The Occupiers' Liability Act 1957 and
- The Occupiers' Liability Act 1984.

The OLA 1957 identifies the liability of occupiers of property (including land) and their duty of care toward lawful visitors and parties to contract entering the property. The Act aims to impose the duty to ensure that visitors are reasonably safe from injury when entering the property for the purpose for which they were invited. This targets liability for injuries arising out of the state of a premises or things that have not been done.

Have you ever noticed how keen supermarkets are when there is a spillage on the floor and so a risk of slipping to customers? This is down to the OLA, as it is their responsibility to ensure that the premises are safe for customers (i.e. visitors) to walk about the shop (the purpose for which they were invited).

Occupiers who clearly display a sign that states that they 'do not accept liability for personal injury' will find this to be invalid under this Act as they cannot dismiss their duty of care. However, warning signs are allowed for example, 'Mind Your Head', 'Slippery Surface', 'Shallow Water' and 'Danger'.

This Act also included liability to landlords and their duty of care to maintain the property to prevent possible injury to a non-tenant, as they are deemed to be in charge of the property.

The OLA 1984 extends the 1957 Act to include persons other than visitors, for example trespassers.

The impact of these acts, therefore, imposes a legal requirement on occupiers of property to ensure that their property is in a safe maintained condition so as to hopefully eliminate the risk of injury to any person to the property. This legal requirement exists whether work is being undertaken on the property or not, but, of course, the risk of injury to others is much greater if significant works are being undertaken on property. Suitable precautions will need to be undertaken such as clearly visible warning signs, lighting, locked gates and fencing and so on.

3 Arranging and organising the work

Questions addressed in this chapter:

How much planning and organisation is involved before I can build?
Why should I use a professional designer?
Should I build it myself?
Should I manage the contract myself?
Do I need a formal contract?
How do I ensure that I am not employing a 'cowboy builder'?
Are there any insurance considerations?
How do I make sure that I am not paying for poor work?
What happens if my contractor carries out faulty work and then goes out of business?

USE OF PROFESSIONAL DESIGNERS

Achieving a satisfactory design is one thing, achieving an effective design that meets your needs and maximises the space available is quite another.

The use of a professional designer may appear an extravagance you can ill afford, particularly if you think that you can produce adequate drawings yourself. However, there are numerous reasons for making use of the expertise of a competent, professional designer:

- Such experts have years of experience that will be used to benefit the design of your property improvement and will ensure you achieve your intended design needs.
- They will also have useful contacts in the industry that could range from trusted contractors to suppliers and include other professional relationships, such as with local authority Planning Officers and Building Control Inspectors.
- They should have professional indemnity insurance in place. This insurance should give you peace of mind and security should there be a problem in the future that was related to the design.

It is not just architects who can provide these services, most building surveyors, for example, will also offer a similar provision.

The use of a professional designer should ensure smooth development and construction phases, as their expertise and experience should eliminate potential problems long before they occur.

Extending and Improving Your Home: An Introduction, First Edition. M.J. Billington and C. Gibbs.
© 2012 M. J. Billington and C. Gibbs. Published 2012 by Blackwell Publishing Ltd.

Even the simplest of improvements requires adequate planning and organisation to ensure the project is carried out with a minimum of fuss. The planning phase does just involve getting your design drawings completed but also requires you to obtain the necessary planning permission and building regulation approval before you can commence. Even from an optimistic viewpoint, these legislative necessities can take at least 8 weeks. However, should any amendments be required to the design following unsuccessful applications, an increase in this period can subsequently occur.

Once acquired, you have several years (see the separate sections on Planning & Building Regulations) before commencing. Therefore, it is best to start the planning stage as soon as possible to ensure everything is in place for when you are ready to build. It is worth remembering that planning permissions and building regulation approvals will pass on to subsequent owners. So, even if you change your mind and decide to sell your property without doing the improvement, the prospect of having the necessary permissions in place can increase the value of your property on the open market.

The amount of information produced during the design stage of your development should be as detailed as possible. The more the information and detailing that can be provided on the drawings and specification, the more likely will be the chances of minimising the scope for disparity and confusion. This confusion can lead to disputes and disagreements that will inevitably cost you time, money and stress. It is therefore important to make decisions early on in the process and stick to them. Think carefully about the level of finish you desire, and the quality and type of materials. Choose your fittings (light switches etc.) and any fitted units. Spend sufficient time to think about these decisions and do your best to stick to them as choices such as these, if ill-conceived, can cause problems much further down the line.

GETTING A BUILDER OR DOING IT YOURSELF?

You may be confident in your DIY abilities having carried out minor alteration and repair work at home, but constructing a complete extension, for example, is considerably much more demanding in several areas:

- Planning
- Organisation
- Technical content
- Physical construction.

The advantages of using a competent builder are similar to that of the professional designer previously discussed (experience, training, contacts and insurance). In addition, a builder will also carry public liability Insurance, which will insure you against damage done to your property and possessions whilst work is being carried out.

Another factor to take into account is time. Do you have enough of your own time to devote to this project? Are you willing to? Is your family willing for you to? Many a home project has strained even the strongest of relationships!

All contractors work for a minimum of 5 days (some even for weekends to ensure efficient completion of the project), can you do the same?

Of course, another important factor not yet mentioned is cost. It is a widely accepted belief that using a contractor will probably use more of your hard-earned cash than you would like to part with. However, a contractor can also save you money:

- Their expertise will minimise the likelihood of technical hiccups and logistical nightmares occurring – *can you ensure the same?*
- They have the plant and equipment readily available for use – *how many extra tools and so on would you have to purchase or hire?*
- They can often acquire preferential rates from suppliers owing to their buying power and trade relations. The same could also be said for any subcontractors that may be required.
- Specialists such as electrical contractors and heating engineers, for example, are often subcontractors to the main contractor. Contractors have preferential subcontractors they use who will have a proven track record and may also obtain privileged rates for this continual 'partnership' – *what do you know about the subcontractors you would potentially employ?*

Choosing a builder you can trust

It is most important to be able to choose a builder that you have trust in. Most people choose a builder by recommendation. Perhaps a neighbour or relative has had work done and can give you first-hand knowledge of their experiences, and, of course, you can view their completed work and see if it is of the required quality. As mentioned above, if you use a professional designer, they will undoubtedly work with a number of contractors that they have known for many years and can vouch for. After all, they would soon stop using a contractor if they found them to be incompetent or unnecessarily expensive.

Except for very minor, low-value work, it is usual to obtain at least three quotations so that you can get a good idea if the prices are competitive. Of course, all contractors must be quoting on the same basis, so you must provide them all with the same technical information, whether this is simply drawings of the proposal or, better still, drawings and specification. Again, a professional designer will be able to manage the preparation of these documents for you and also handle any negotiations over price with the successful contractor.

There are a number of other ways that the quality of contractors can be judged if none of the above suggestions are applicable and you want to manage the job yourself. In order of preference, these can be divided into government-supported schemes, membership of trade bodies and commercial vetting schemes.

Government-supported schemes

These are 'not-for-profit' schemes established to enhance the reputation of the construction industry and the companies that work in it.

Trustmark

This initiative was set up by a former government as an attempt to overcome the problem of the 'cowboy builder'. The original scheme was called Quality Mark and proved to be too expensive to be run by government, so was passed over to the private sector to run. It is open to virtually all building trades and is administered by TrustMark. Any organisation that wants to run a TrustMark scheme must apply to TrustMark for approval. At present there are over 30 scheme operators working in the repair, maintenance and improvement sectors of the construction industry. These include trade associations and local government trading standards teams as well as independent scheme operators. These schemes are approved to carry the TrustMark logo and recruit reputable and trustworthy tradesmen. All of these schemes are audited annually by TrustMark to ensure processes, standards and complaint procedures are being maintained.

Scheme operators must ensure the following:

- Each firm's technical skills are independently checked through regular on-site inspections, as well as checks on their trading record and financial status.
- Firms sign up to a code of practice that includes insurance, good health and safety practices and customer care.
- The firm's quality of work, trading practices and customer satisfaction processes are checked and monitored.
- Firms offer an insurance-backed warranty.
- Deposit protection insurance is available for consumers in the event that a firm should cease trading.

Additionally, using a TrustMark registered contractor will have the following benefits:

- Firms will be able to tell you about any Building Regulations you must comply with and may also be able to provide appropriate certificates (if members of a Competent Persons Scheme);
- If you have a problem or disagreement with the firm, there will be a clear and user-friendly complaints procedure to help resolve the issue.

It is important to check that when employing a TrustMark tradesman, they are 'licensed' for all the trades/work you are asking them to carry out. This can be done by looking on the TrustMark web site and searching under the 'trade' then 'their company name' and finally 'more information' – you will then see a list of the trades that the firm is licensed for under the protection and standards offered by TrustMark.

TrustMark can be located at www.trustmark.org.uk.

Competent Person Schemes

Competent Person Schemes (CPS) were introduced by the Government to allow individuals and enterprises to self-certify that their work complies with the Building Regulations as an alternative to submitting a building notice or using an approved inspector.

The principles of self-certification are based on giving people who are competent in their field the ability to self-certify that their work complies with the Building Regulations without the need to submit a building notice, thus incurring local authority inspections or fees. The government hopes that moving towards self-certification will significantly enhance compliance with the requirements of the Building Regulations, reduce costs for firms joining recognised schemes and promote training and competence within the industry. It should also help tackle the problem of 'cowboy builders' and assist local authorities with enforcement of the Building Regulations.

Detailed information on Competent Person Schemes is given in Chapter 2.

Membership of trade bodies

Many contractors of all sizes are members of trade bodies or federations appropriate to their particular trade. In fact, some trade federations run TrustMark Schemes and are also Competent Person Scheme operators in their own right. One such example is the National Federation of Roofing Contractors (www.nfrc.co.uk). Some trade federations, such as the NFRC, actively ensure that members offer high standards of workmanship and sound business practice through a strict code of practice and independent vetting procedure, including site inspections and adhering to TrustMark standards. Additionally, all NFRC trade members hold comprehensive insurance cover and insurance-backed warranties are also available if required by the client.

Unfortunately, not all trade organisations provide such comprehensive checking and monitoring services and it is worth checking a trade federation's web site to see the type of checks that are carried out on their members.

Commercial vetting schemes

A glance in Yellow Pages will reveal that many contractors are members of various 'competence checking' organisations that can be identified by their logos in the advertisements. No doubt, while many of these are reputable, it should be noted that they are commercial profit-making organisations charging member companies an annual fee to belong. They often do not carry out comprehensive checks on their member's technical abilities or financial status and rely instead on testimonials from 'satisfied customers' in order to 'approve' their listed members.

This practice of collecting logos for advertisements is known as 'badging' in certain circles and it should be noted that not all badges are of the same value.

COST IMPLICATIONS

It is not unusual for a project to go over the anticipated cost. This can happen for several reasons, the most common of which are

- Change of specification by client;
- Unforeseen works (often found during excavation);
- Insufficient information at costing/estimate stage and
- Additional works requested by client.

As with anything else you purchase, the key to getting 'value for money' is to shop around. It may not always be the case to go for the cheapest, however. Check out the history or track record of a company. Ask them about previous jobs it has carried out and ask if you could speak to its clients. The construction industry thrives on 'word of mouth' and recommendations to acquire more work and improved reputation; it is not necessarily just a question of buying a bigger advertisement.

Depending on the size of your project, a deposit may be required to show commitment to the contractor. However, be careful and ensure that this is a deposit toward the final contract sum and not money to buy materials. A contractor should have sufficient funds or accounts with suppliers to enable purchase of materials until at least the first stage payment (if the job allows stage payments).

CONTRACTUAL ARRANGEMENTS

There are many ways in which a project can be arranged in terms of the contractual relationships. You may at first think that it would be easiest if you were the 'nucleus' of all that goes on, but when you look at it from a contractual viewpoint it can be the most confusing and most frightening option (Figure 3.1).

Just think that each line is a separate contractual relationship that requires organising, developing, agreeing and monitoring. Therefore, the more in control you want to be, the much more complex the role becomes. So long you are prepared for the time commitment and stress associated with this arrangement, it can work quite effectively, remembering that organising individual subcontractors to work seamlessly to your timetable can be a minefield on its own.

As can be seen from Figure 3.2, by releasing some of the control and appointing solely a main contractor, the appointment, control and responsibility of subcontractors can be released to the main contractor.

The more detailed the drawings and specification, the less the risk exists for any contractual disputes or claims. It is, therefore, worthwhile devoting sufficient time to the design stage of your project and not be in too much of a rush to get it started. Insufficient time spent here could escalate into much more time wasted and costs incurred further

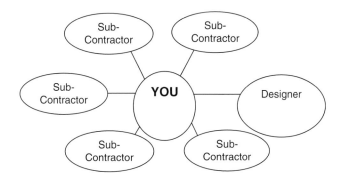

Figure 3.1   Managing everything yourself.

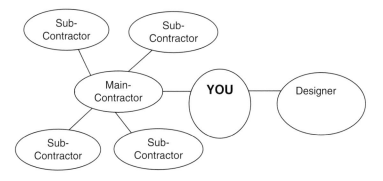

Figure 3.2   Using a main contractor.

down the line. In Chapter 1, the stages involved in a project from inception to completion were described. These go from A to L. It is somewhat salutary to realise that the construction stage does not start until K!

While there are standard forms of contract available for small projects (e.g. minor works), very often the contract consists of a letter, drawings and specifications together with the contractors tender or estimate.

Standard forms of contract may appear very complex and unnecessary, but these contracts have been professionally produced and developed over many years with both parties to the contract in mind. Therefore, they can offer much more protection but you must ensure you fully understand the workings of a construction project and the roles and responsibilities of those involved.

Knowing your own limits is paramount, not only in this area of endeavour but also for all things in life. People have undergone years of training to develop a career in a specialism that could help you. Understandably, they are not charities and, of course, will require payment for their services. It is our view that this is money well spent.

AVAILABILITY OF GRANTS

Over the years there have been a great many different grant schemes, operated primarily by both central and local government. Grant schemes are often politically motivated (often with the best of intentions) and are usually set up by a government when there is a perceived need to alter the way in which the populace lives or behaves.

Examples of past schemes that are no longer available include the following:

- The home improvement grants of the 1960s, aimed at ensuring that all dwellings were substantially free from damp, had inside toilets, bathrooms and, later, central heating.
- Grants to enable people living in Council-owned Precast Reinforced Concrete (PRC) homes to have these made structurally sound and weathertight so that they could raise a mortgage and purchase the freehold.
- The recent boiler scrappage grant scheme aimed at assisting people to replace old-fashioned inefficient boilers.
- The nationwide microgeneration grant scheme for installing PV panels and so on.
- The Home Energy Efficiency Scheme for Wales.

As initiatives vanish, they are generally replaced by others that are politically expedient. At present, with global warming being the big issue and with the government under pressure to live up to its Kyoto protocol commitments, various energy saving grants are available. Some of these are discussed in the following sections.

Warm Front grant

The Warm Front scheme provides heating and insulation improvements to households on certain income-related benefits living in properties that are poorly insulated and/or do not have a working central heating system.

Qualifying households can get improvements worth up to £3500 (£6000 where oil central heating and other alternative technologies are recommended).

Grants are available for improvements such as

- Loft insulation
- Draughtproofing
- Cavity wall insulation
- Hot water tank insulation
- Gas, electric, liquid petroleum gas or oil heating
- Glass-fronted fire – the Warm Front scheme can convert your solid-fuel open fire to a glass-fronted fire.

Nothing is payable by the householder as long as the work does not cost more than the grant available. If the cost of the work is more than the grant available, the householder must make a contribution to enable work to go ahead. Work will not start unless the householder is willing and able to pay the difference.

Additionally, the property must also be poorly insulated and/or not have a working central heating system.

Local council grants

You may be able to get energy efficiency grants and discounts from your local council. To qualify you may need to be a pensioner and/or claim certain benefits. To find out if you are entitled to anything, contact your local council.

Energy suppliers

The Government's Energy Efficiency Commitment (EEC) has been replaced by the Carbon Emission Reduction Target (CERT). This means that energy suppliers with a certain number of customers operating in Great Britain are obliged to achieve targets for improving home energy efficiency.

The suppliers, therefore, provide a range of offers which significantly reduce the cost of installing energy efficiency measures. Also, you can take up offers from any of the energy companies, regardless of who supplies your gas and electricity.

The best source of information for all energy saving grants is the Energy Saving Trust (www.energysavingtrust.org.uk).

Grants and loans for work to listed buildings

Listed buildings are defined and discussed in Chapter 2. Grants are available in certain circumstances both from English Heritage (the Historic Buildings and Monuments Commission for England) and from local authorities. They are always at the discretion of the body giving them and listing does not give any automatic entitlement to a grant.

Repairs grants are available for buildings of 'outstanding architectural or historic interest' and are made towards reroofing, treating dry rot and other structural repairs, but not normally towards decoration or works of regular maintenance. Owners have to show that they would not be able to complete the work without financial help and are usually asked to supply details of assets and income to substantiate their application.

Applications should be made to English Heritage (the Historic Buildings and Monuments Commission for England) (www.english-heritage.org.uk).

Local authorities may make grants for any building of architectural or historic interest and are not restricted to outstanding buildings or even to listed buildings. Grants may be made by County and District Councils (in London by the London Borough Councils) and enquiries should be addressed to the appropriate local authority.

Conservation area grants are available for buildings in conservation areas and are made for works which will make a significant contribution to the 'preservation or enhancement of the character or appearance of a conservation area'. Usually the grants are administered by local authorities in conjunction with English Heritage. Your local authority should be contacted for details of any schemes available in your area.

Finally, some listed buildings enjoy a more favourable position on the payment of Value Added Tax (VAT) on works than do unlisted buildings. Repairs and alterations to unlisted buildings are subject to VAT at the standard rate, but alterations to listed buildings that are designed as dwellings or used for qualifying residential or non-business charity purposes, together with those that are being converted to such use, are not subject to VAT, as long as the work is done by a VAT registered builder and with listed building consent.

INSURANCES FOR CONTRACT AND AFTER

It is obvious that sometimes things go wrong with a building project. Perhaps you have commissioned a professional designer to prepare the drawings and details or a structural engineer to do structural calculations and they make a mistake or omit to consider something that later leads to defective work or materials. Equally, your builder may fail to control the work properly or use substandard materials and you may find that your new roof leaks or your extension develops cracks and suffers from settlement.

It can be quite difficult to apportion blame in these circumstances. Since a leaking roof may be due to poor design or defective workmanship and materials, you may find that the designer blames the builder and vice versa. This can leave you in the unpleasant position of having to take one or the other (or both) to court to recover costs and get the work put right, meaning that you have to risk yet more money with no certainty that you will be successful.

There are a number of remedies that are open to you to mitigate these circumstances. Above we mentioned the use of standard building contracts. These set out the responsibilities of each of the parties to the contract (whether it is a contract with a builder or a designer) and, therefore, indicate the paths that can be followed in disputes.

Much is often made of the fact that a particular builder may offer a 'guarantee' for his work. These usually vary from 5 to 10 years, whereby the contractor agrees to put right any defects that are proved to be his fault within the terms and conditions of the guarantee. The homeowner may find such a guarantee quite hard to enforce, especially if the terms and conditions are unduly restrictive and, in any case, the guarantee will not be enforceable if the contractor has gone out of business.

Fortunately, the much-maligned insurance industry has come up with a number of products that address many of these issues and some of these products are described below.

Professional indemnity insurance

Common wisdom in today's litigious world has it that if something goes wrong, then somebody must be to blame, so therefore that somebody will have to pay. Professional indemnity insurance (PII) is a form of liability insurance whereby cover is provided for the financial consequences of professional negligence where there has been a breach of professional duty by way of neglect, error or omission. In addition, an indemnity is provided in respect of the legal and other costs and expenses incurred in the defence of any claim.

If you employ a professional, for example, to carry out design or do structural calculations, then you rely on expertise which they purport to have and you do not. Such a professional cannot ignore an allegation of professional negligence. The allegation must be defended or admitted and, as a result, there will be cost implications in either case.

Any professional person providing advice, design, specifications, supervision and so on owes a duty of care to their client and third parties, and it does not matter whether they do this for a fee or provide their services for nothing (as is sometimes the case when a family member does work for a close relative etc.).

The test for the duty of care owed is generally whether or not they have exercised reasonable skill and care in the discharge of the services provided. If a professional fails to exercise this duty (i.e. is negligent), they may be liable for losses incurred by their client, and/or third parties. Taking into account the operation of the current legal system, even proving innocence can be very costly.

A professional purchases PII for their own protection. The cover is not for the benefit of the professional's client, although far too often it is not seen that way. Clients cannot claim directly against the professional indemnity insurance cover carried by the professionals – they must prove liability first, a process which can be time consuming, expensive and uncertain.

Although liability still has to be proved, the client employing a professional covered by PII at least knows that if he can prove liability, then there will be the means to pay the claim. Nothing is worse than pursuing a person through the courts only to find that when success is ultimately achieved, they have no money.

It should be noted that the major professional institutions in the United Kingdom that cover architects, chartered surveyors, chartered engineers and so on make it a condition of membership that their members hold PII.

Insurance against the risk of accidents affecting property or persons

Some types of insurance are aimed at ensuring that where accidents occur (and building work can be a dangerous activity), there is sufficient money available to repair any damage or assist the recovery of injured persons. The main insurances in this area are public liability insurance and employers' liability insurance.

Public liability insurance

Statute laws impose a 'duty of care' upon certain people, including employers, suppliers of goods and services and owners of property. If this duty of care is breached, then the individual or business may be liable for damages. This is particularly pertinent in the case of contractors who visit the homes of their clients to carry out potentially dangerous activities such as replacing windows or renewing electric wiring. Public liability insurance covers an individual or business for potential liability to third parties that experience personal injury or property damage.

In most cases, public liability insurance is a voluntary measure. In practice, however, a contractor may find that customers will not deal with them unless they have adequate cover and most good contractors will provide such cover, especially if they are members of a reputable trade federation where the provision of public liability insurance will be a condition of membership (see details on the National Federation of Roofing Contractors in an earlier section).

The main advantage to a contractor in having public liability cover is that the legal costs associated with defending a claim or paying damages will be covered.

From the client's viewpoint, they will know that if, through the contractor's acts or omissions, they suffer injury or their property is damaged, then there will be insurance cover in place to ensure that due recompense is received.

Employers' liability insurance

Employers are responsible for the health and safety of their employees while they are at work. A contractor's employees may be injured at work or they, or former employees, may become ill as a result of their work while employed by the contractor. They might try to claim compensation from their employer if they believe that he is responsible. The Employers' Liability (Compulsory Insurance) Act 1969 ensures that at least a minimum level of insurance cover against any such claims is provided.

Employers' liability insurance enables the cost of compensation to be met for employees' injuries or illness whether they are caused on- or off-site. However, any injuries and illness relating to motor accidents that occur while employees are working for the contractor will most probably be covered separately by the contractor's motor insurance.

Employer's liability insurance is compulsory for most companies in the private sector, but is not mandatory for sole traders.

Insurances for site works

It has been mentioned above that a professional may carry professional indemnity insurance to cover any alleged negligent acts, errors or omissions. It should be noted that the client will still have to prove liability for negligence. Therefore, the insurance protects the professional and not the client. Mention has also been made of the fact that a contractor's guarantee for the works carried out will not be worth the paper it is written on if the contractor goes into liquidation. There is also the question of any deposit the client may be asked to pay before the work starts. Unfortunately, it is known for unscrupulous persons to take a number of client's deposits and then do a vanishing act with their client's money.

Again, the insurance industry has come up with products designed to cover these eventualities in the form of Bondpay, latent defects insurance (LDI) and insurance-backed guarantees (IBGs). The following information has been provided by kind permission of

Quality Assured National Warranties (which as Warranty Services Ltd is a company authorised and regulated by the Financial Services Authority) (www.qanw.co.uk).

Bondpay

Bondpay is a payment system whereby the client makes stage payments into a trust account which effectively commits their money to the contractor without the risk of the contractor not providing the contracted work. At each stage the work is completed to satisfaction, the payment is released directly to the contractor's bank account and the next stage payment is paid into the trust. This not only ensures that the client's money is never at risk but also gives the contractor confidence that the client will pay.

Latent defects insurance

A latent defect is one which remains undiscovered at the date of practical completion but manifests itself during the period of insurance by way of actual physical damage. Latent defects insurance is a form of insurance that meets the cost of rectifying defects in materials or workmanship and errors or omissions in design or specification which are not apparent when the job is finished but which come to light later.

Latent defects insurance protection is usually arranged for the construction of new commercial and industrial buildings, refurbishments and certain elements, such as cladding (including external insulation), curtain walling, glazing, rendering, asphalting, roofing and car park decking, in isolation or in combination. However, it can also be considered for extensions and conversions to both commercial and domestic buildings.

The emphasis on quality is supported by long-term insurance security that maintains its value over time. Ten, 15 and 20 year single-premium index-linked policies are available for most projects. The minimum contract value is £20 000.

The insurance covers

- Insolvency
- Design
- Workmanship defects
- Material defects.

The advantage of latent defects insurance is that it will pay for the work to be corrected where a defect occurs that is covered by the terms and conditions of the policy whether or not the contractor is still in business. If necessary, the insurer still has the right to pursue the faulty party (designer, builder or supplier of the defective materials) through the courts for recovery of the costs involved in correcting the works without this affecting the end client. The insurance can be purchased by the client or by the contractor.

Insurance-backed guarantees

Normally IBGs are used for contracts up to the £20 000 threshold, subject to the status of the builder. IBGs are available for virtually all trades and categories of work that they might do (a full list is provided on the QANW web site mentioned above), although the terms and cost will vary. An IBG is a low cost, long-term insurance policy, which provides valuable protection for consumers when undertaking home improvements projects. The principle of an IBG is to honour the terms of the written guarantee originally issued by the installing contractor, where that contractor has ceased to trade as defined within the policy document and is therefore unable to satisfy claims against that guarantee. If Bondpay is not used, the IBG can also protect any deposit paid by the client if the contractor goes out of business before starting the job.

Therefore, an IBG is an insurance policy that covers the contractor's written guarantee in the event that the contractor ceases to trade (the concept is defined in the policy). If a defect occurs and the contractor is still trading, they would be expected to correct the defect under the terms of his guarantee. Usually, the contractor arranges their own IBG facility with the insurer and provides it as an added benefit to customers who use them. If this is not the case, the customer can still request that an IBG be provided by the contractor who would need to contact an insurer such as QANW to obtain the cover.

FURTHER INFORMATION

Further information can be obtained from

http://www.communities.gov.uk/
http://www.nfrc.co.uk
http://www.qanw.co.uk
http://www.english-heritage.org.uk
http://www.energysavingtrust.org.uk
http://www.qanw.co.uk.

4 Issues affecting the property as a whole

Questions addressed in this chapter:

Are there structural risks to undertaking work on my property?
Is getting the building watertight important?
Why should I consider maintenance of my property?

STRUCTURAL STABILITY

When undertaking any refurbishment work it is important to ensure that the structural integrity of the building is not affected or impaired in any way. This may appear quite obvious but if you are not sure how all the elements of a building work together and not just in isolation you may cause structural problems during or after the work.

- *Are you aware of the impact of enlarging openings (doors and windows) in a building?*

An opening in a structural element (e.g. a wall) is a weakness and, therefore, too many openings or openings that are too large can have an adverse effect on the structural stability of that wall and possibly on other elements of the building that rely on that wall for support.

Structural stability of buildings is covered by Part A in Schedule 1 of the Building Regulations 2010. Part A is supported for the giving of practical guidance by Approved Document A – Structure (Chapter 2 covers the legal status of Approved Documents).

From a general Building Regulation viewpoint the requirement is for buildings (including extensions and alterations) to be constructed so that all dead, imposed and wind loads are sustained and transmitted to the ground safely. Dead loads are calculated by adding together the weight of the materials that make up the building and these are of a known constant amount. Imposed loads (sometimes called live loads) are those that result from the use of the building and are variable from hour to hour, day to day; for example furniture, snow on the roof and people in the building. A structural engineer will use a figure for the imposed loads derived from statistical data that averages out the maximum expected load on the building.

When you make an opening in an existing wall or enlarge an existing opening you change the load path, that is the path taken by the loads on the building on their way to the foundations and, ultimately, to the ground. For example, if you form a new opening in an

Extending and Improving Your Home: An Introduction, First Edition. M.J. Billington and C. Gibbs.
© 2012 M. J. Billington and C. Gibbs. Published 2012 by Blackwell Publishing Ltd.

existing wall and it is too close to a corner of the building it may be that the loads transmitted by the new lintel that you have installed will increase the load on the corner to such an extent that the supporting wall collapses.

Openings or recesses in a wall should not be placed in such a manner as to impair the stability of any part of it or to adversely affect the lateral restraint offered to the wall by a buttressing or return wall. Adequate support for the superstructure should be provided over every opening and recess.

As a general rule, any opening or recess in a wall should be flanked on each side by a length of wall equal to at least one sixth of the width of the opening or recess, in order to provide the required stability. Accordingly, the minimum length of wall between two openings or recesses should not be less than one sixth of the *combined* width of the two openings or recesses.

However, where long span roofs or floors bear onto a wall containing openings or recesses it may be necessary to increase the width of the flanking portions of wall. Table 8 of Section 1C of Approved Document A contains factors that enable this to be done.

Where several openings and/or recesses are formed in a wall, their total width should, at any level, be not more than two thirds of the length of the wall at that level and should not in any case exceed 3 m in total.

These requirements are illustrated in Figure 4.1.

■ *Are you aware of the impact of forming chases in a wall in order to provide new or altered service runs?*

The depth of vertical chases should not be more than one third the thickness of the wall or, in a cavity wall, one third the thickness of the leaf concerned. Depth of horizontal chases should be not more than one sixth the thickness of the wall or leaf. Chases should not be placed in such a manner as to impair the stability of the wall, particularly where hollow blocks are used.

■ *Do you know how to cut holes or form notches in floor and roof timbers so that new services, such as electric wiring, conduits and pipes for plumbing can be installed safely without weakening the floor or roof?*

Notches and holes in floor and roof joists should comply with the details shown in Figure 4.2. However, no notches or holes should be cut in rafters except for birds-mouths at supports. The rafter may be birdsmouthed to a depth of up to one third of its depth. Notches and holes should not be cut in purlins or binders unless checked by a competent person.

■ *Are you aware of the work that can be undertaken to internal walls?*

Internal walls in buildings can fall into three categories, namely:

(1) Non-load-bearing partitions
(2) Load-bearing partitions
(3) Compartment walls

(See also Chapter 9, Internal Walls.)

Non-load-bearing partition walls are generally lightweight in construction and are a means of dividing up the floor space in a building into rooms. They are built directly off

Approved Document A, Table 10

Nature of roof span	Maximum roof span [m]	Minimum thickness of wall inner leaf [mm]	Span of floor is parallel to wall	Span of timber floor into wall max 4.5 m	Span of timber floor into wall max 6.0 m	Span of concrete floor into wall max 4.5 m	Span of concrete floor into wall max 6.0 m
				Value of factor 'X'			
Roof span parallel to wall	Not applicable	100	6	6	6	6	6
		90	6	6	6	6	5
Timber roof spans into wall	9	100	6	6	5	4	3
		90	6	4	4	3	3

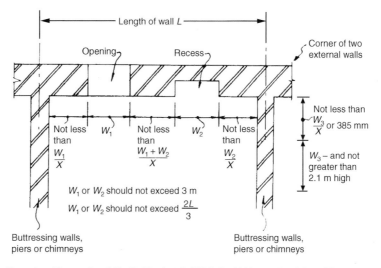

Note: value of X comes from Table 10 of Section 1C of AD A1/2 which is reproduced above OR it may be
given the value 6 provided the compressive strength of the blocks or bricks (or cavity wall loaded leaf)
is not less than 7 N/mm².

Figure 4.1   Openings and recesses in walls.

the floor and do not provide a means of support to any other element of the structure. It is generally safe to remove these walls. However, professional advice should be sought to ensure that no adaptations have been made to the wall that could cause problems and to determine what services, if any, could be located in the wall.

Load-bearing partition walls are similar to non-load-bearing walls but they should have their own foundation. These partitions may provide support to the floor and/or roof, so removal of these walls may not be carried out without some remedial work to provide the support that the wall is doing. Usually it is necessary to insert a steel, concrete or timber beam in place before the entire wall is removed, and this must be done extremely carefully or the structure above will be compromised. It is not unknown for incompetent builders or ignorant householders to take out a wall without providing adequate temporary support and then have to cope with cracking in the structure above or, at worse, the total collapse of half the house. This sort of alteration is covered by the Building Regulations

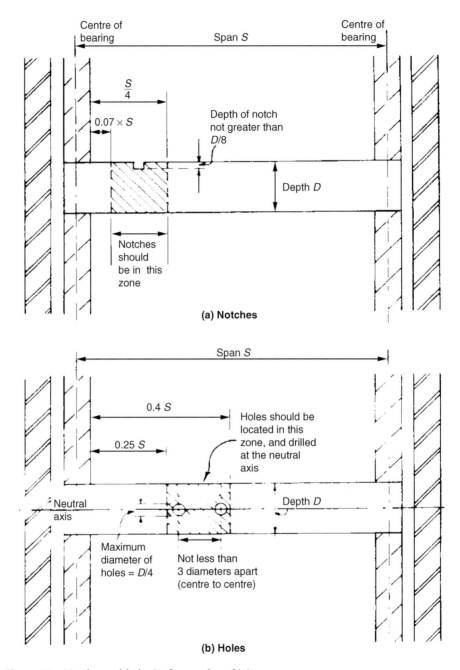

Figure 4.2   Notches and holes in floor and roof joists.

and requires an application to be made. The application would need to be supported by structural calculations prepared by a suitably qualified person (such as a chartered structural engineer). If the inserted beam is steel it would need to be encased in non-combustible construction (such as plasterboard) to provide the necessary 30 min fire resistance. Interestingly, although a steel beam is non-combustible (except at very high temperatures) in a fire it quickly loses its strength and buckles or collapses. By contrast, a large section timber beam (say, $200 \times 150 \, \text{mm}^2$) is combustible but has adequate fire resistance and does not need to be protected. This is because when wood burns it chars and the resultant surface charcoal provides a degree of fire resistance sufficient to satisfy the regulations.

Compartment walls can be either non-load-bearing or load-bearing. However, they also provide a means of protection in case of fire. For example, they may be protecting a staircase to provide a safe means of escape. Removal of these walls could, therefore, cause a danger to the occupants of the property and could also contravene Building Regulations.

■ *Can I form a new opening in the floor for a staircase?*

Forming a new opening in a suspended upper floor for a staircase is possible provided that the construction of the opening is formed appropriately. Forming the opening will necessitate the shortening of several joists, so care must be taken to ensure that the lateral restraint given to the walls by the floor is not removed. (Also see Chapter 7, Floors.)

Figure 4.3 illustrates that if a new stairwell was to be formed in this upper floor at least five joists may have to be trimmed. These trimmed joists were providing lateral restraint to the external wall, therefore provision must be made to ensure restraint is maintained. On this subject Approved Document A makes it clear that in certain circumstances where it may be necessary to interrupt the continuity of lateral support for a wall for a stairway or similar structure this is permitted provided certain precautions are taken:

■ The opening extends for a distance not exceeding 3 m measured parallel to the supported wall.
■ If the connection between wall and floor is provided by means of mild steel anchors, these should be spaced closer than 2 m on either side of the opening, so as to result in the same number of anchors being used as if there were no opening.

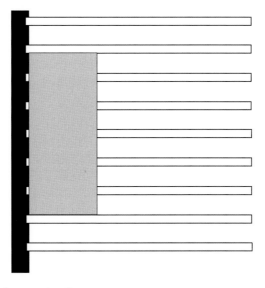

Figure 4.3 Proposed new stairwell.

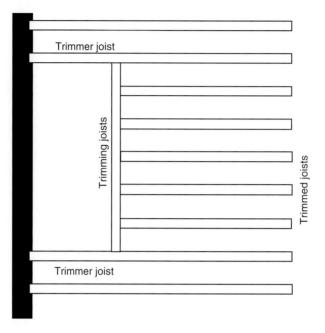

Figure 4.4 Trimming of a new opening in a floor.

- Other forms of connection (i.e. than mild steel anchors) should be provided through out the length of each part of the wall on either side of the opening.
- There should be no other interruption of lateral support.

The trimmed joists must be supported. This is achieved by using a trimming joist. As shown in Figure 4.4, the trimming joist is supported by the first full joist on either side of the opening. These are termed 'trimmer joists'.

Both the trimming joist and trimmer joists should be thicker in size, as they are taking additional loads from the trimmed joists. It is common practice for the trimming and trimmers to be the same cross-sectional size as the other joists but they will be 'doubled-up' (i.e. two joists side by side) and bolted together.

- *Can I take out the fireplace and chimney breast?*

Back in the late 1960s and 1970s this used to be quite a popular way of increasing the usable space inside a house. It was often done in conjunction with the removal of a wall between the front and back rooms in a house to provide a 'through lounge'. The advent of central heating made people think that they no longer needed that old, dirty open fire. Now that we are more energy conscious and appreciate the impact that a real fire can contribute to a living space, many people are actually rebuilding chimneys and fireplaces in their houses and, of course, modern efficient wood-burning stoves have contributed to this trend.

If you want to remove a fireplace or chimney breast, similar considerations will apply as for the removal of a load-bearing wall. Always seek the advice of a competent structural engineer, as support systems for chimneys and chimney breasts vary and it is most important to consider the existing load paths carefully and ensure that changed load paths are adequately catered for. Do not forget that if you have a chimney and chimney breast attached to a party wall then Party Wall Act considerations will apply (Chapter 2) and your neighbours consent will be required.

WEATHERPROOFING AND DAMP-PROOFING THE ENVELOPE

A major milestone in any development project, irrespective of the size and type of project, is the day on which the project becomes sealed from the elements of the weather, or in more common construction terms 'watertight'.

Until this point it is generally only external works and the building shell (or envelope) that can progress. The building envelope incorporates the external walls and roof. Whilst it also incorporates the windows and doors, it is not uncommon for these to be formed with temporary wooden frames and polythene sheeting during actual construction to exclude the weather. This prevents the actual finished windows and doors from damage during the rest of the works. These are then installed in place of the temporary frames after the heavy work is completed.

The importance of this watertight milestone is related to several factors, for example:

- This may also be a milestone activity for the release of money from any lending body.
- The remainder of the works internally can proceed unaffected by the vagaries of the weather.
- Planning the project now becomes more complex as numerous activities can progress simultaneously, for example first fix carpentry and joinery, first fix plumbing, first fix electrical, external works and external finishes. First fixing (sometimes called carcassing) is when the electric and other cables and water and other service pipes are installed. It refers to all the services that must go in before the walls are covered with plaster or plasterboard.
- By making the building watertight and excluding the weather the building structure is already starting its 'drying-out' period. This is significant as a new project, even an extension, can take over a year to fully reach its normal moisture content depending on the methods of construction and materials used.

The exclusion of moisture from the property is important not only to leave a dry internal environment but to reduce the amount of movement that will occur from the drying out process. An increase in moisture in porous building materials causes expansion and as the material dries out contraction (or shrinkage) occurs, so cracking could be evident if this is not minimised. This can be said of building materials such as blockwork, brickwork, mortars, plaster, concrete and timber, which can severely twist, warp, crack and split if drying out occurs too quickly.

Attempts to reduce the adverse effects of rapid drying out have led to different construction methods being adopted, for example:

- Timber frame construction uses much less water in its construction, so less drying out is expected.
- Dry lining a property (lining the internal surface with plasterboard) as opposed to a traditional sand/cement render and plaster reduces the amount of moisture being added to the building.

■ Beam and block flooring can also reduce the amount of moisture, particularly if a timber decking is used instead of a sand/cement topping (called a screed).

All of these aspects not only have an effect on the amount of moisture being put into the building but can also speed up the process of construction, so the watertight milestone can be reached sooner. This will, of course, mean that the building is open to the weather for a shorter period and so less 'additional' water from rain, snow and so on will be added to the structure.

Unfortunately, these processes may be more expensive in terms of materials costs than more traditional techniques but, as they involve less labour on site, the cost may work out to be similar to traditional methods if you have to pay for that labour. In addition, a quicker construction time may also result in a quicker completion time, so reducing the amount of any additional borrowing, interest payments, storage, insurances, equipment hire and so on. Take all factors into account when considering the cost effectiveness of a product or building element rather than just the initial purchase price of the materials or product.

It is also important to take care of the building materials when they are not being used. Good housekeeping on site is imperative, not only to limit the amount of waste materials on site but to limit the amount of additional moisture absorbed by the stored materials. Keeping materials off the ground (i.e. not in direct contact), covered and allowing free movement of air around them will keep them in a stable condition, limit shrinkage from moisture movement and reduce the effect of ground contaminants such as salts (e.g. efflorescence).

This phase of construction and the additional items identified play an important role towards the successful completion of a project and show that what may appear to be small items in relation to the overall project, can have much greater influences than first thought.

MAINTENANCE

Over time the materials that make up your home may deteriorate if they are not looked after (i.e. maintained). There are different types of maintenance that must be considered.

■ Planned, preventative maintenance.
■ Responsive maintenance.

Planned, preventative maintenance is essential to the life of the materials used in the building. If the materials are affected then the structural integrity of the building, and in severe cases the health of the occupants, can be affected.

Planned maintenance can be regarded as the everyday maintenance that should be carried out to preserve and protect. This can be quite simple to achieve if it is done regularly and appropriately. Long term it can also save you money, as you will not have to replace materials or elements of the building and it will make your home more saleable if it is kept in good condition.

Examples of planned maintenance include the following:

■ Cleaning of rainwater gutters, downpipes and gullies twice a year.
■ Clearing air vents (e.g. underfloor ventilation to timber floors).

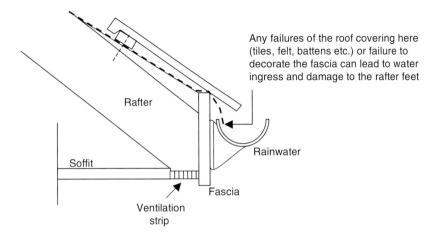

Any failures of the roof covering here (tiles, felt, battens etc.) or failure to decorate the fascia can lead to water ingress and damage to the rafter feet

Rafter

Soffit

Rainwater

Fascia

Ventilation strip

Figure 4.5 Damage under the eaves caused by lack of maintenance.

- Clearing of plant growth and so on from walls and chimneys.
- Redecoration of external timber joinery (fascia, soffit, windows etc.) at least every 5 years.
- Replacing cracked or loose render.
- Repainting of rendered external walls at least every 5 years.
- Testing smoke and carbon monoxide detectors.
- Annual service of gas and/or oil appliances and central heating system.
- Electrical test at least every 10 years.

Items such as these reduce the risk of long-term defects that may occur due to lack of essential maintenance. For example, regular redecoration of the external joinery (fascia, soffits, windows etc.) will be considerably more cost effective than having to replace them sometime owing to rotting of the timber. In addition, while this timber has been rotting owing to lack of maintenance it could also be causing further problems, such as rotting of the feet of rafters and damage to internal finishes.

As shown in Figure 4.5, lack of maintenance to fascia or rainwater gutters could lead directly to damage to the foot of the rafter.

Responsive maintenance is work that must be carried out to a property owing to damage generally caused by freak weather or poor planned maintenance. Clearly it would be almost impossible to plan against freak weather conditions (e.g. gale force winds, flooding etc.). However, well executed planned maintenance could reduce the effect the weather would have and, therefore, still reduce the potential cost of remedial works.

FURTHER INFORMATION

- Building Regulations – Approved Document A.

5 Site survey and investigation

Questions addressed in this chapter:

What is the importance of a site and subsoil investigation?
What happens if I have large trees in my garden and I want to extend my house?
Why is it important to have an accurate measured survey of my property?

To some 'the site' may simply be a piece of land that is going to have something built on it. However, there are a multitude of considerations to take into account, each of which could have a significant impact on the project if not treated with respect. Cutting corners here could affect the overall design and increase the time and costs of your project, so a full understanding of your site and its peculiarities is highly important.

SITE INVESTIGATION

In all but the simplest forms of building alteration and improvement, it is essential that someone carries out a thorough examination of the area or location of the proposed changes. This is referred to as a site investigation and all professional designers and surveyors are trained in the required techniques. This may not be true of the average builder, although the best contractors, who may also be qualified and experienced tradesmen, will understand the importance of a good site investigation.

This importance can be understood by considering the following factors, which illustrate the perils of going into a major alteration programme unprepared.

■ *If you are building a house extension, do you know anything about the original house foundations – their construction, depth and the type of soil they are founded on?*
It is unfortunate if having budgeted for a project you find that when the contractor arrives on site and digs out the trench for the foundations, they find that they must go much deeper than assumed and will want to increase their estimate.
■ *Do you know where the underground services, such as drains, water pipes and electricity cables are located?*
It is all too common for the contractor to find that an unexpected drain run interferes with a proposed foundation trench, meaning that either the foundations have to be re-designed or the drain has to be re-sited – again with cost implications.
■ *Do you know what effect that nice big oak tree in your garden will have on your proposals?*
Many species of trees can have serious effects on the design of buildings and extensions. This is discussed in more detail later in this chapter.

Extending and Improving Your Home: An Introduction, First Edition. M.J. Billington and C. Gibbs.
© 2012 M. J. Billington and C. Gibbs. Published 2012 by Blackwell Publishing Ltd.

■ *Do you know if you are in a radon gas affected area or you are near to a landfill site (perhaps now not used but still potentially problematic)?*

If you alter or extend a house which might be affected by radon or landfill gas, then special precautions are needed under building regulations to mitigate the effects of these potentially life-threatening contaminants.

■ *If you are carrying out internal or external alterations such as the removal of a wall or the installation of a new window, do you know what form of construction you are cutting into or removing?*

Houses vary enormously in their forms of construction, especially with regard to walls, roofs and floors. Illustrated in this book are some of the more common forms that you can adopt; they are based on sound building principles. You should always find out the form of construction and whether it is load bearing or not before you contemplate alterations, otherwise you may find that your alteration becomes structurally unsound or allows the penetration of dampness.

■ *Do you know where the internal service runs are for electricity, gas, hot and cold water and heating services?*

Removing an internal wall or digging up a floor and finding a water, gas or electricity main supply can not only be expensive but also potentially dangerous.

■ *Do you know what sort of damp-proof course your house has?*

Damaging an existing damp-proof course can mean that you may suffer from rising and penetrating dampness after you have completed your alterations.

■ *Is your house free of the effects of wood-boring beetle attack and wet and dry rot?*

A thorough survey will reveal evidence of these risks to the timber construction in your house and will allow you to take advantage of the works you are planning to budget for and correct these defects at the same time.

■ *Have you had an accurate measured survey done of your house or at least the areas that you are proposing to alter or extend?*

The measured survey is key to any house alteration or improvement and a good surveyor will pick up many of the items referred to above when making their measurements. If done badly, it can have the effect of making it necessary to alter or amend your plans when work starts and discrepancies are discovered. The principles of measured survey are covered in more detail later in this chapter.

Site investigation process

The full site investigation process would not usually be needed where it is intended to extend or alter an existing dwelling. However, some notes on the process are included to show the sort of material that is available and can be collected. For the scale of work covered by this book, a lesser survey would be sufficient, although the principles remain the same.

A classic site investigation starts with a **desk study**. This involves the collection of as much information as possible about a site from sources such as geological maps, Ordnance Survey maps, air photographs, geological books, civil engineering magazines, mining records, reports of previous site investigations and from local knowledge such as that held by the local authority building control department. Your local reference library can also be a source of invaluable information especially if it has a local history section.

At the same time, this will enable you to find out about the history of your house, especially if it is old or interesting.

The second phase of a site investigation is referred to by surveyors as the **walk-over survey**. The objective of the walk-over survey is to check and make additions to the information already collected during the desk study and includes visiting the site and its surrounding area and covering it carefully on foot. During this process, the surveyor would ask questions of local authorities, local inhabitants and people working in the area, such as builders, electricity and gas workers, in order to obtain the benefit of their local knowledge before the production of a structured report based on the information gathered.

The report produced would cover the following topics:

- Topography (i.e. the nature of the slope of the land etc.) of the site, including slope angles
- Expected ground water conditions
- Geological setting (what types of subsoils could be expected)
- Location, size and species of any trees, shrubs or hedges, either at present or in the past, with notes on dates when removed
- The presence of any aggressive chemicals found in the ground and groundwater
- Probability of pre-existing slope instability (i.e. the tendency of some types of soil, such as clay, to slip down steep slopes when exposed to heavy rain etc.)
- Position of existing and demolished structures
- Possible extent and dates of mineral extraction and mining in the area
- Evidence for the existence of made ground on site (such as rubble fill)
- Possible locations for structures
- Likely types and loading of structures
- Access for excavators and boring/drilling rigs during ground investigation
- Position of services (e.g. electricity, water, gas, telephone, sewers).

The final part of the site investigation would be the subsoil investigation, which is discussed next.

SUBSOIL INVESTIGATION

It is essential that all building loads are carried safely to the foundations and then distributed safely to the subsoil. Therefore, it is vital that the nature of the soil is determined in order that appropriate foundation selection is identified.

Trial pits (holes dug at intervals around the site of the proposed works) and boreholes (holes formed using a small portable drilling rig) can be used to investigate the subsoil conditions. It is important that these investigations are carried out at critical locations, for example, on the foundation line of the external walls, near corners of proposed extensions and near the line of any internal load bearing walls. All too often investigations have been carried out but not located in relation to the proposed building and, if the ground conditions vary across the site, large problems requiring expensive solutions can occur.

It is also important to identify if the proposal is in an area that suffers from localised issues. An example of this is shrinkable clay soils. Clay soils occur in Southern England roughly south of a line between the River Severn and the Wash. Shrinkable clay soils occur widely and particularly in areas such as London and the South East of England. These soils are exceptional in their ability to swell and shrink excessively depending on their water content which varies seasonally.

As mentioned above, the local building control office can often provide valuable information on neighbouring ground conditions. as well as being able to tell you if extensions have already been carried out at your own house and, if so, what kind of ground conditions were encountered.

SIGNIFICANCE AND PROXIMITY OF SITE BOUNDARIES

It has been mentioned above that it is important to know and measure the position of site boundaries. Very often these are marked by walls, fences or hedges. However, one of the commonest forms of disputes between neighbours is over the position of a boundary between properties. Most people think that their site boundary will be marked on the plans that accompany the deeds of their property as lodged with the Land Registry. Most people will be disappointed! Our experience is that the plans are often

■ Not to scale;
■ Of such a small scale that it is not possible to position the boundaries accurately enough;
■ Of such poor reproduction that they are almost impossible to read and
■ Without any dimensions.

Only England and Wales followed the example of Jersey in the Channel Islands. There, all boundaries have to be marked with permanent boundary stones actually on the site.

A further complication that arises is that of ownership of the wall, fence, hedge and so on that runs along the boundary. There are no hard and fast rules about this. Sometimes the fence posts may be on one or other side of the rails; sometimes, in the case of trellis fencing, the posts will be positioned on the centreline of the fence. With walls, there may or may not be piers that are thicker than the general construction of the wall and it is often the case that the piers are sited on the wall owners side of the boundary but this is not always the case. Hedges are even more problematic. They may originally have been planted on the boundary line but over time will have grown to such an extent that the boundary may no longer be discernible. The plants themselves often give rise to disputes, especially with vigorous varieties such as the Leyland Cypress × *Cupressocyparis leylandii* (syn. *Callitropsis × leylandii*). This is often referred to as just Leylandii and it is a fast-growing evergreen tree much used in horticulture, primarily for hedges and screens. Even on sites with relatively poor soil, plants have been known to grow to heights of 15 m in 16 years. Their rapid, thick growth means they are sometimes used to enforce privacy, but such use can result in disputes with neighbours whose own property becomes overshadowed.

Apart from questions of ownership, the importance of establishing the boundary line cannot be overstated for the following reasons:

■ There are controls under the Party Wall Act (Chapter 2) when it is proposed to carry out works on or near a boundary.
■ Often it is not possible under the Planning Laws to build on the front or side of a house or to build nearer than 7 m to a rear boundary (Chapter 2).
■ Building Regulations and Planning Laws impose restrictions on the type, amount and area of windows (and doors under the Building Regulations) in a new extension facing a side boundary.

PRESENCE AND SIGNIFICANCE OF TREES

In the section on subsoil investigation above, specific reference was made to the problems of building on shrinkable clay soils. Where shrinkable clay soils are present on a site, the presence of mature trees can exacerbate the tendency of the soil to cause heave (i.e. swelling) and subsidence (i.e. a lowering of the building due to shrinkage of the ground below it) and this may lead to damage to services, floor slabs and oversite concrete. In shrinkable clay soils, soil shrinkage caused by the removal of water by trees can lead to foundation failure due to subsidence. Conversely, soil swelling can occur following tree removal as the soil moisture levels recover. This can lead to foundation heave. So it is not an answer to simply fell the tree, unless this is done slowly over a couple of years so that the ground can recover naturally before the building work is done. The downside of this is that it is possible that the tree is protected by a Tree Preservation Order, in which case felling or lopping it is a criminal offence, and it may be that the tree is a particularly fine specimen and you do not want to fell it anyway.

Obviously, the simplest way to avoid tree root problems is to site the building work at a safe distance from the offending tree so that the tree cannot influence the moisture content of the soil under the building. If this approach is adopted, then the general rule is that the building work should be at least as far from the tree as its expected maximum height. This would mean that for a mature oak tree, the building would have to be about 20 m away. This would mean that for most normal sites it would be virtually impossible to extend the building at all. Clearly, this is not necessarily the best solution.

The other possible solutions are as follows:

■ Use ordinary concrete strip foundations and deepen them to get below the level at which tree roots cause problems (the zone of desiccation).
■ Design special foundations.

Deepen the foundations

The basis of design is to ensure that the foundations are deep enough so that they will not be affected by tree roots. The trees themselves do not necessarily have to be large varieties,

Table 5.1 Foundation distances and depths for different tree species.

Distance from building (m)												
1	2	4	6	8	10	12	14	16	18	20	22	24
Species Foundation depth (m)												
Oak 2.70	2.60	2.45	2.30	2.15	1.95	1.80	1.65	1.45	1.30	1.10	0.95	0.90
Poplar 2.70	2.60	2.50	2.40	2.25	2.15	2.00	1.90	1.70	1.60	1.45	1.30	1.20
Willow 2.70	2.55	2.50	2.40	1.95	1.85	1.55	1.40	1.20	0.95			
Hawthorn 2.55	2.40	2.10	1.75	1.45	1.00							
Leylandii 2.60	2.40	2.00	1.75	1.20	0.90							
Cedar 1.65	1.50	1.20	0.90									
Fir 1.65	1.50	1.20	0.90									
Pine 1.65	1.50	1.20	0.90									
Spruce 1.65	1.50	1.20	0.90									
Chestnut 1.75	1.65	1.50	1.40	1.30	1.15	1.00						
Ash 1.75	1.65	1.50	1.40	1.30	1.15	1.00						
Lime 1.75	1.65	1.50	1.40	1.30	1.15	1.00						
Sycamore 1.75	1.65	1.50	1.40	1.30	1.15	1.00						
Pear 1.65	1.60	1.30	1.05									
Cherry 1.75	1.65	1.50	1.30	1.15	1.00							
Alder 1.75	1.65	1.50	1.30	1.15	1.00							
Maple 1.75	1.65	1.50	1.30	1.15	1.00							
Beech 1.75	1.65	1.50	1.30	1.15	1.00							
Plum 1.65	1.50	1.20	0.90									
Laurel 1.65	1.50	1.20	0.90									
Apple 1.65	1.50	1.20	0.90									
Laburnum 1.15	1.05	0.90										
Birch 1.20	1.10	0.95										
Holly 1.20	1.10	0.95										
Magnolia 1.15	1.00											
Mulberry 1.15	1.00											

such as Oak or Beech; even small trees, such as Plum or Apple, can have an effect on foundations.

Table 5.1 gives the minimum depth required for the foundations for some different tree species depending on how far the foundations are from the trees.

Special foundation designs

Special foundations include reinforced concrete rafts, or short-bored or driven pile and beam solutions. These types of foundations must be designed by a structural engineer and so would be more expensive to design, However, modern short-bored pile systems have now been developed to such an extent that they are at least as competitive as traditional deep strip foundations, mainly because the alternative of digging a very deep trench for traditional foundations would be costly anyway. With short-bored piles and ground beams, the excavation is reduced to a minimum and they are less dependent on good weather for their completion.

MEASURED SURVEY

Simply put, a measured survey is the act of physically measuring and recording the dimensions of an existing building and/or that area of the site of a building where an extension is to be sited (including the position and proximity of boundaries). A measured survey is required if scale drawings need to be produced for planning permission, building regulations, contractor tendering, design purposes and to detail the works to be undertaken.

By taking accurate measurements, scale drawings of the existing building can be produced either manually or by using specific computer software such as AutoCAD.

Equipment is basic for carrying out a measured survey and includes the following:

- Sketching/writing equipment: pencil, pen, paper and clipboard
- Measuring equipment: 5 m steel tape, 30 m tape or laser distance measure
- Camera: to record features and take general photographs for records or reference

Both horizontal and vertical measurements need to be recorded. It is always better to take too many measurements rather than the basic minimum, as additional measurements will serve as a means of checking the used data when producing the drawings.

The types of drawings to be produced include

- Existing and proposed floor plans;
- Existing and proposed elevations (i.e. drawings of the sides of the house and extension);
- Construction details;
- Site plan and
- Section through building and possibly the site depending on the work being undertaken.

The type of measurements that need to be recorded include

- Room widths, lengths, heights and diagonals (to check that the walls are not out of square);
- Window positions including sill height, height of window and type of window;
- Door position, including height, width, hinge side and direction of swing, and
- Thickness of walls (internal and external).

Depending on the nature of the work to be undertaken, it may be necessary to record additional items, such as the location of

- Boiler, radiators and so on;
- Meters (gas, electric and water);
- Services (electric runs and socket outlets, pipe runs, above and below ground drainage including rainwater goods).

Figures 5.1 and 5.2 show typical survey sketches and site notes for, in this case, a possible single-storey extension to a two-storey building. Note that only the part of the building being altered has been measured. It would also be necessary to measure the part of the garden adjacent to these alterations, including the distance to the boundary.

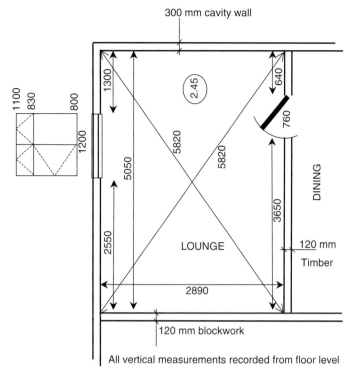

Figure 5.1 Site notes floor plan. (The figure shown circled in the drawing is the floor to ceiling height.)

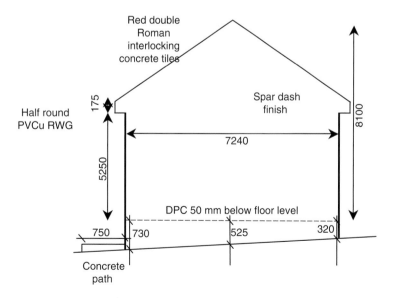

Figure 5.2 Site notes elevation.

The accuracy of any drawings produced, any calculation of proposed works (stair design, roof design etc.) and the estimate of the proposed works (materials and labour) is determined by the accuracy of the site notes and recordings made during the measured survey site visit. Therefore, it must be stressed that you need to take care, take time, take too many measurements, take photographs and make plenty of large fully annotated sketches.

FURTHER INFORMATION

- Building Regulations – Approved Document C

6 Foundations

Questions addressed in this chapter:

What types of foundation are there?
What factors determine the choice of foundation?
How do the Building Regulations affect the choice of foundation?
How can services passing through foundations be dealt with?

INTRODUCTION

The purpose of the foundation is

- To carry and distribute all the loads of the building safely to the subsoil below;
- For an extension to a building, to distribute all the loads in such a way as not to upset the structural stability of the building that is being extended;
- To take account of the nature of the subsoil and not be disrupted by swelling, shrinking or freezing of the subsoil;
- To be designed to resist possible landslip or subsidence to which the site might be prone and
- To provide a firm level base from which to commence any wall or frame construction.

In the past, especially during the Victorian period, it was common for foundations to be formed in diminishing widths of brickwork laid in the bottom of a trench, a technique known as brick spread footings. If you are extending an old house, you may find this type of foundation construction when you dig down to form your extension foundations. Additionally, you may find that the existing foundations are not laid as deep as they would be put today and as would be required by current Building Regulations.

It is very important that the link between the new and old foundations is formed so that the new building does not adversely affect the stability of the original building. There are various techniques that can be used to form this junction and there are two main principles, either of which can be followed depending on the circumstances:

(1) Where the existing ground is firm and not subject to seasonal movement (i.e. is not shrinkable clay), it is normal to bond the two foundations firmly together. The new foundation is placed at the correct depth required by Building Regulations (discussed later) and, if necessary, is run under the existing foundation for a distance of 150–225 mm, thereby wrapping round the existing foundation at the point of junction for the full width of the new foundation trench. The thickness of concrete below the existing foundation should be at least 150 mm.

Extending and Improving Your Home: An Introduction, First Edition. M.J. Billington and C. Gibbs.
© 2012 M. J. Billington and C. Gibbs. Published 2012 by Blackwell Publishing Ltd.

(2) Where the existing ground may be subject to seasonal movement, it is usual to treat the existing and new buildings separately and to not tie them tightly together. This allows the extension to move vertically over time without disturbing the existing foundations. This technique requires special detailing and expertise; so if this is encountered when trial pits are excavated during the site investigation (Chapter 5), the services of a structural engineer should always be sought.

FOUNDATION TYPES

Modern construction techniques use concrete as the principal material for forming foundations. A concrete foundation is placed on prepared ground at the base of the proposed walls.

With the exception of a framed building (which is not common construction for domestic cases), the foundation type chosen is determined by the soil type and ground conditions.

Strip foundations

A strip foundation is the most common and, for most site conditions, the most economic type of foundation (Figure 6.1). This type of foundation can be used on level or sloping sites where the foundation would be formed by a series of 'steps' (Figure 6.2).

A trench is excavated to a specific depth that is determined by the nature of the subsoil. However, the minimum recommended depth is 760 mm to eliminate the risk of frost heave of the ground.

The stepped foundation should be constructed for the following reasons:

- The upper level overlaps the lower level by twice the height of the step, by the thickness of the foundation or 300 mm, whichever is the greater.
- The height of a step is not greater than the thickness of the foundation.

Additionally, the foundation strip should project beyond the faces of any pier, buttress or chimney forming part of a wall by at least as much as it projects beyond the face of the wall proper. The amount of projection needed and, therefore, the ultimate width of the foundation trench will depend on the type of subsoil encountered. As a general rule, the projection should not be less than the thickness of the concrete used in the foundation unless the foundation strip is reinforced with steel reinforcing bars, and this would require the services of a structural engineer to do the calculation. Some typical strip foundation designs are shown in Figure 6.3; these illustrate the Building Regulation solutions for two types of soil.

Therefore, the width of the trench is determined by the foundation width, which, in turn, is determined by the load of the structure resting on it, the load-bearing capacity and type of ground and, to some extent, the thickness of the wall.

Table 10 in Approved Document A (Table 6.1) gives the minimum width of strip footing (the old word for foundations but still commonly used in the building trade) for different types of ground and imposed loads. The table also gives details of simple tests that can be carried out to test the strength of the ground. With regard to the loadings, which are given in

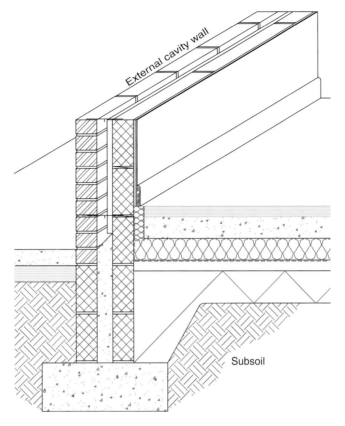

Figure 6.1 Strip foundations.

kilo Newtons per linear metre (kN/m) of foundation, the figures relate approximately to the forms of construction shown in Table 6.2.

Once the foundation trench width has been decided, the concrete will be placed level in this trench to a minimum practical thickness of 225 mm. Whilst this thickness may seem excessive for a single-storey building or extension, it provides for the possibility of a second storey in the future without having to provide very expensive underpinning works

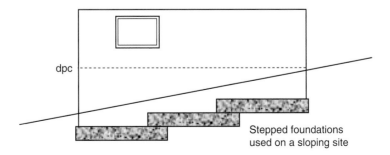

Figure 6.2 Stepped foundations.

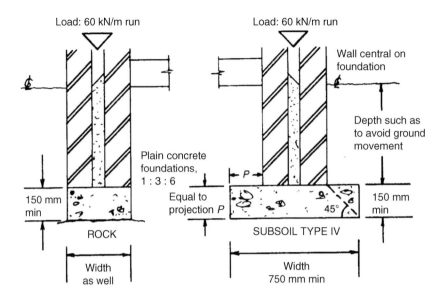

Figure 6.3   Typical foundation designs to comply with Building Regulations.

to the foundations. In fact, Approved Document A of the Building Regulations permits a minimum thickness of 150 mm but as using exactly this thickness would require very accurate excavation and bottoming up (i.e. smoothing out) of the foundation trench, the minimum practical thickness is 225 mm as mentioned above.

It is important that the proposed wall is located centrally on this foundation to ensure no overturning of the foundation occurs (Figure 6.4).

Service ducts can be provided through the new wall during the construction up to damp-proof course level by leaving appropriate holes in the wall at the necessary location with lintels placed above these openings (Figure 6.5).

In the case of the passage of a drain through a foundation, special precautions may be necessary due to the size of the hole required and the possibility that a conduit might be formed under the ground that could allow the passage of vermin (rats and mice) or ground gases into the dwelling. There is also a risk that settlement of the building may cause pipes to fracture, with consequential blockages and leakage. In the past it was common practice to require pipes (which were rigid jointed) to be encased in concrete. Since the development of flexible pipe systems, it has become essential to maintain this flexibility in order that any slight settlement of the building will not cause pipe fracture.

Where a drain is built into a structure (e.g. a wall, foundation, ground beam, drainage inspection chamber, manhole etc.), suitable measures should be taken to prevent damage or misalignment. The following solutions are possible:

■ The wall may be supported on lintels over the pipe. A clearance of 50 mm should be provided round the pipe perimeter and this gap should be masked on both sides of the wall with rigid sheet material to prevent the ingress of fill or vermin. The void should be filled with a compressible sealant to prevent ingress of gas.

Table 6.1 Minimum width of strip footings.

Type of Ground (Including engineered fill)	Condition of ground	Field test Applicable	Total load of load-bearing walling not more than (kN/linear metre)					
			20	30	40	50	60	70
			Minimum width of strip foundation (mm)					
I Rock	Not Inferior to sandstone, limestone or firm chalk	Requires at least a pneumatic or other mechanically operated pick for excavation	In each case equal to the width of wall					
II Gravel or Sand	Medium dense	Requires pick for excavation. Wooden peg 50 mm square in cross section hard to drive beyond 150 mm	250	300	400	500	600	650
III Clay Sandy Clay	Stiff Stiff	Can be indented slightly by thumb	250	300	400	500	600	650
IV Clay	Firm	Thumb makes impression easily.	300	350	450	600	750	850
Sandy Clay V Sand	Firm Loose	Can be excavated with a spade.	400	600				
Silty Sand	Loose	Wooden peg 50 mm square in cross section can be easily driven.	400	600	Note			
Clayey Sand	Loose				Foundations on soil types V and VI do not fall within the provisions of this section if the total load exceeds 30 kN/m.			
VI Silt	Soft	Finger pushed in up to 10 mn	450	650				
Clay Sandy clay Clay or Silt	Soft Soft Soft							
VII Silt Clay Sandy Clay Clay or Silt	Very Soft Very Soft Very Soft Very Soft	Finger easily pushed in up to 25 mm	Refer to specialist advice					

The table is applicable only within the strict terms of the criteria described within it.

Table 6.2 Loading in relation to forms of construction.

Load (kN/m)	Type of construction resting on foundation
20–30	Single-storey construction in lightweight and traditional construction, respectively
40–50	Two-storey construction in lightweight and traditional construction, respectively
60–70	Three-storey construction in lightweight and traditional construction, respectively

Any building heavier than 70 kN/m is not covered by the table and the foundation will have to be designed by a structural engineer.

■ A length of pipe may be built in to the wall with its joints not more than 150 mm from each face. Rocker pipes not exceeding 600 mm in length should then be connected to each end of the pipe using flexible joints (Figure 6.6).

Trench fill or deep strip foundations

An alternative type of strip foundation is a trench fill (Figure 6.7). This type of foundation requires the trench to be filled with concrete to just below ground level. This eliminates the often difficult aspect of constructing walls in the narrow spaces in the trench, obviates the need for temporary trench support and so speeds up the ground works element of the construction. The additional cost of the mass concrete is offset by the lower materials and labour costs of the substructure wall construction below damp-proof course, although there may be a problem as well as an expense in getting rid of the

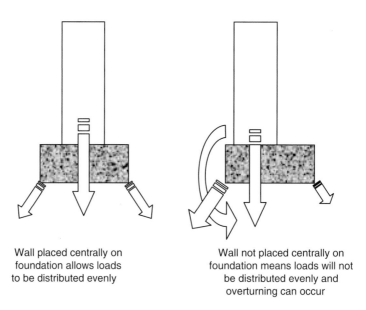

Wall placed centrally on foundation allows loads to be distributed evenly

Wall not placed centrally on foundation means loads will not be distributed evenly and overturning can occur

Figure 6.4　Overturning of foundations.

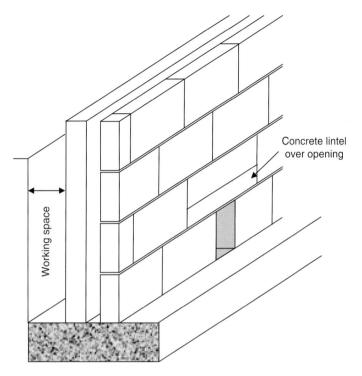

Concrete lintel
over opening

Working space

Figure 6.5 Provisioning for service ducts.

excess excavated soil unless this can be relocated on the site to form new landscaping features. The width of the foundation is designed in exactly the same way as for traditional strip foundations.

Wide strip foundations

Where the subsoil is less suitable, it may be possible to use a wide strip foundation. This is similar to a traditional strip foundation but is wider to enable the building loads to be distributed over a larger area. Structural calculations are required for this type of foundation, as mesh reinforcement is needed at its base to ensure the wall does not 'push' through the concrete and so eliminate the benefits of the additional width (Figure 6.8).

The proposed wall construction is carried out in the same way as a strip foundation. However, there will be more working space in the trench owing to the extra width, which will also require additional backfill material.

The costs of excavation and foundation construction can vary greatly, so it is vital that adequate soil investigation must be carried out by the use of trial pits or bore holes at specific locations to ensure the subsoil type is accurately identified. Should investigations determine that the subsoil is unsuitable for the previously mentioned foundation types, then it may be necessary to provide a more complex foundation, such as a raft foundation or short-bored piles and ground beams.

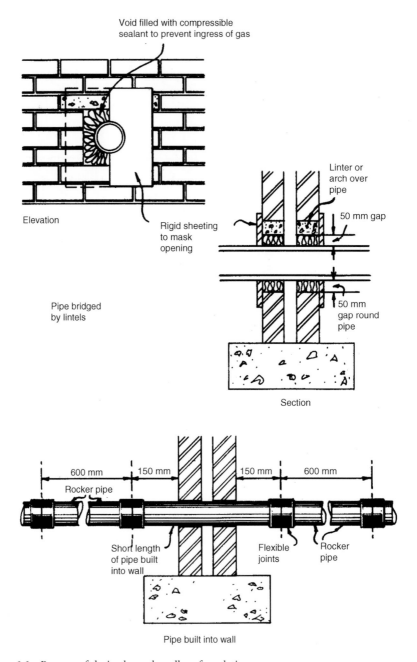

Figure 6.6   Passage of drain through wall or foundation.

Reinforced raft foundations

A raft foundation incorporates a reinforced thickened edge beam and ground floor in one monolithic construction (Figure 6.9). This has the benefit of safely distributing the building loads over the whole of the building area. This type of foundation must

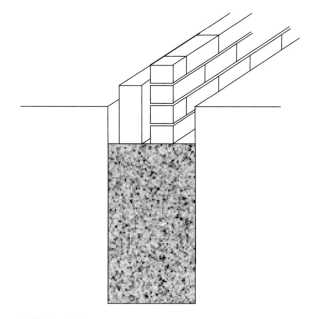

Figure 6.7 Trench fill foundations.

be designed by a structural engineer and can include some complex reinforcement. It is also important that the building is designed symmetrically over the raft to ensure that the loads are distributed evenly over the whole area and are not greater on one side, as subsidence or leaning of the finished building may occur at some time in the future.

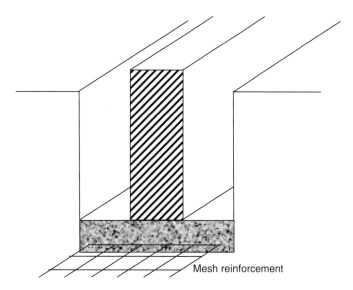

Mesh reinforcement

Figure 6.8 Wide strip foundations.

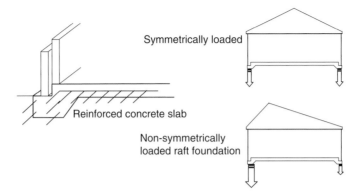

Figure 6.9   Reinforced raft foundations.

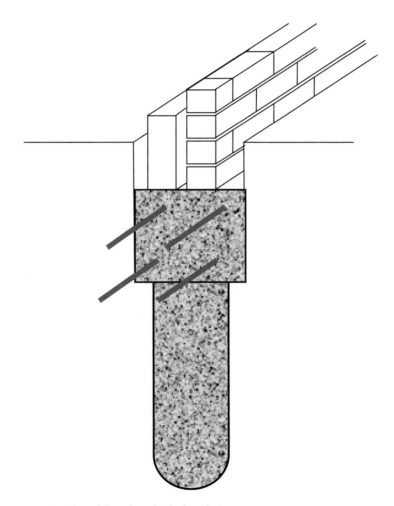

Figure 6.10   Principles of short-bored pile foundations.

Short-bored pile foundations

Short-bored pile foundations are most often used in shrinkable clay soils where large trees are present and excavating below the zone of desiccation would be expensive due to the depth that they would require (Chapter 5).

Piles are like concrete columns but in this case they are placed in the ground to support the building rather than being placed above ground to support the roof. They can be made of precast concrete or steel and then driven or screwed into the ground or, more usually where the ground is composed of clay, a hole will be augured (drilled) into the ground by machine and then filled with a pre-constructed reinforcement cage and finally concreted.

Piles that reach no more than 4 m below ground level are termed short-bored piles and they are usually between 2 and 4 m in length and 250–350 mm in diameter. The principles of short-bored pile foundations are shown in Figure 6.10.

One advantage of using bored piles is that the soil that is expelled during the drilling process can be analysed to see if there are any variations in soil type below the ground. This allows the depth of the piles to be adjusted if the soil strata are seen to vary in depth.

The piles are sited at corners and at the intersections of load-bearing walls and also at intermediate positions, so that the loads of the structure above are evenly distributed via the ground beams to the ground. The ground beams are simply reinforced strips of concrete cast into fairly shallow trenches near to the ground surface.

Such foundations will, of course, require the services of specialist subcontractors who can carry out the necessary subsoil investigations and prepare the structural designs needed for building control.

This is a brief description of the most common types of foundations that are met in domestic extension work. Other types of foundation are available but these are normally designed and installed by specialist contractors, so are not covered within this book.

FURTHER INFORMATION

Reference used in this chapter:

- The Building Regulations 2010 – Approved Document A

Further information can be obtained from

http://www.roger-bullivant.co.uk

7 Ground and upper floors

Questions addressed in this chapter:

How can I construct a ground floor?
How can I construct an upper floor?
Are there any new materials and methods in use for these constructions?
What Building Regulations apply to these construction areas?

GROUND FLOORS

There are a number of designs that can be used to construct a ground floor for the refurbishment of a dwelling or an extension. Each type has its own merits, but whichever type is chosen they must all satisfy the same requirements. These include

- Preventing the passage of moisture from the ground;
- Providing a level and visually acceptable wearing surface that performs the function demanded by the room in which it is situated;
- Withstanding the loads placed upon it (imposed and dead) and
- Providing a level of thermal performance.

The general performance of ground floors is also covered by the Building Regulations in Part C and in Approved Document C. Accordingly, ground floors should be designed and constructed so that

- The passage of moisture to the upper surface of the floor is resisted;
- They will not be adversely affected by moisture from the ground;
- The passage of ground gases is resisted. This relates back to Requirement C1 of the Building Regulations where floors in certain localities may need to be constructed to resist the passage of hazardous gases such as methane or radon. The measures shown to resist gas ingress can function as both a gas resistant barrier and damp-proof membrane if properly detailed;
- The structural and thermal performance of the floor is not adversely affected by interstitial condensation. Interstitial condensation is moisture, usually contained in a vapour form within the air inside the building, which condenses within the structure of the floor when it meets a cool surface. This also applies to floors exposed from below and
- Surface condensation and mould growth is not promoted under reasonable occupancy conditions. This applies to all floors (not just those next to the ground).

This guidance is illustrated in Figure 7.1.

Extending and Improving Your Home: An Introduction, First Edition. M.J. Billington and C. Gibbs.
© 2012 M. J. Billington and C. Gibbs. Published 2012 by Blackwell Publishing Ltd.

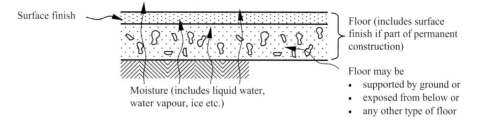

Surface finish

Floor (includes surface finish if part of permanent construction)

Floor may be
• supported by ground or
• exposed from below or
• any other type of floor

Moisture (includes liquid water, water vapour, ice etc.)

Figure 7.1   Floors: general guidance from Building Regulations.

There are three different types of ground floor construction in common use in domestic alterations and extensions:

■ Solid
■ Suspended timber
■ Beam and block

Solid ground floors

A solid ground floor is generally the simplest and most economic form of construction. This floor requires a solid base (i.e. subsoil) for construction as it is in contact with this base and is not suspended in any way. The floor requires the subsoil for support but the direct contact allows for even distribution of loads from the room to the subsoil.

Figure 7.2 illustrates a typical construction detail for a solid ground floor and identifies its main components, which are as follows:

■ Floor screed
■ Oversite concrete
■ Insulation
■ Damp-proof membrane (dpm)
■ Blinding
■ Hardcore
■ Subsoil

The details that follow comply with Building Regulations Approved Document C with regard to the general principles stated above.

The position of the insulation could be changed and placed on top of the oversite concrete. This option prevents the oversite concrete acting as a thermal store, so prompts a quicker response time to the internal environment. This means that the building will warm up more quickly when the heating is first turned on after a cold spell, but it will also cool down more quickly when the heating is turned off. It should, however, be more economical in terms of heating costs.

Choosing this alternative location for the insulation has an effect on the floor screed. As the sand and cement screed is not now in direct contact with the oversite concrete and, therefore, not bonded to it, the thickness of the screed will need to be increased and reinforced with a wire mesh (Figure 7.3). This will prevent the screed from cracking and/or curling at the edges and provide sufficient strength in respect of loadings.

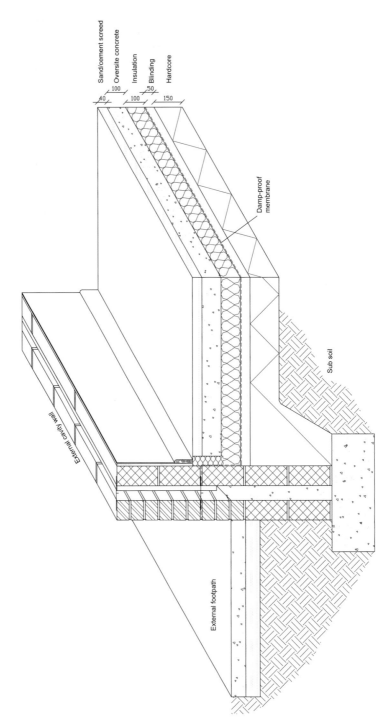

Figure 7.2   Ground floor detail.

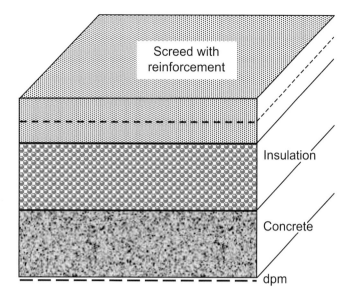

Figure 7.3 Alternative insulation position with reinforced screed.

This alternative location for the insulation also allows an alternative option to replace the floor screed with a timber floor covering such as particleboard, provided that this is moisture resistant (Figure 7.4).

The insulation used is normally flooring grade expanded polystyrene (EPS), as this will not compress significantly under the weight of the floor structure. The thickness of the insulation is dependent on the thermal conductivity of the insulation material and the required U value that needs to be achieved for the whole floor structure. The U value is a measure of the thermal efficiency of the construction and for the ground floor of an extension this should not be greater than 0.22 W/m^{2}K. The actual thickness of the insulation will depend on the material used but for expanded polystyrene the thickness

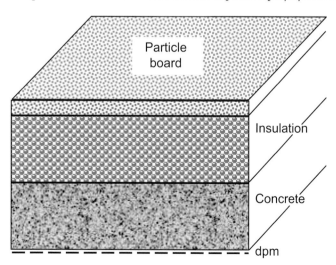

Figure 7.4 Alternative insulation position with particleboard.

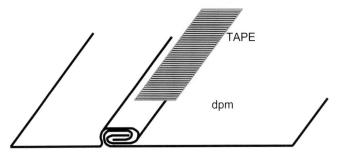

Figure 7.5   Damp-proof membrane lapped joint.

would need to be about 125 mm. For other materials with a better thermal conductivity the thickness could be less.

The oversite concrete base acts as the main structural portion of the ground floor and distributes loads to the hardcore and then subsoil.

Preceding the hardcore is a damp-proof membrane which is normally a thick polythene sheet material of not less than 1200 gauge. This dpm is to prevent the passage of moisture from the subsoil to the interior of the dwelling, so must be lapped with the horizontal damp-proof course (dpc) in the inner leaf of the external walls.

It is preferable that the dpm is laid in one piece to achieve maximum performance. However, if this is not possible, any joints should be lap jointed and taped (Figure 7.5). If there are any protrusions through the dpm (e.g. service ducts or drainage pipes), the dpm should be taped to the duct to ensure it is not displaced during the laying of the concrete and a watertight seal can be maintained. This is particularly important in areas where radon gas is prevalent.

Underneath the dpm is a blinding. This blinding is normally sand. It protects the dpm from being damaged or pierced by any materials contained in the layer of hardcore.

The hardcore material is made up of consolidated/compacted materials, such as broken brick, concrete or ballast. It is important that there is no organic matter such as timber or gypsum plaster, as these will eventually degrade and leave a void beneath the floor that could cause cracking and failure of the oversite concrete base.

The hardcore should be compacted in layers and can be of varying thickness depending on conditions, such as if additional excavation is needed to find a suitable substrata or where the building is on a sloping site. It is recommended that if the hardcore is going to exceed a depth of 600 mm an alternative floor construction should be sought in terms of both economy and minimising the risk of future failure.

One such alternative to a solid ground floor is a suspended timber floor.

Suspended timber ground floor

This type of floor is a traditional type of construction that has been used for over a hundred years. Whilst modern forms of this type of floor are almost identical in design to the older floors, some developments have been included over a period to address potential problems and common defects found in these earlier constructions.

A modern suspended timber floor illustrating a typical method of construction is shown in Figure 7.6. Floor joists support tongued-and-grooved softwood floorboards to form the

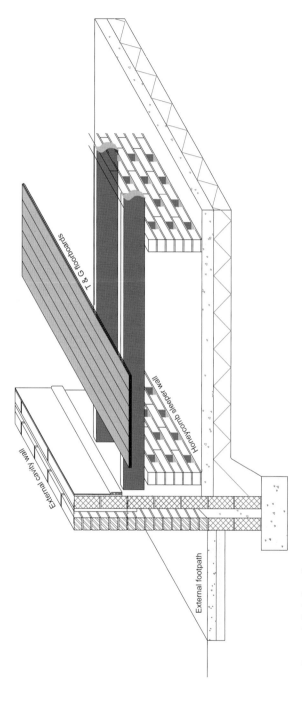

External cavity wall

T & G floorboards

Honeycomb sleeper wall

External footpath

Figure 7.6 Suspended timber floor.

level wearing surface to the interior (sheets of particleboard are often used as a more economic material choice if finish is not important). These floor joists are supported from underneath by sleeper walls built off an oversite concrete base. These sleeper walls are constructed in half brick thick brickwork (i.e. 102.5 mm thick) and are of a honeycomb construction to allow adequate ventilation throughout the floor void.

A horizontal dpc is placed on top of this sleeper wall to prevent the passage of moisture from the ground as a dpm is not required under the oversite concrete. Above the dpc a timber wall plate is placed to provide a level surface to accept the floor joists, spread the load effectively to the sleeper wall and allow a medium for fixing the floor joists.

Insulation is placed between the joists in this type of construction.

The sleeper walls also allow the ends of the floor joists to be supported without being built into the external wall and risk being affected by moisture. This was an inherent problem associated with older type suspended timber floors in Victorian dwellings.

The sleeper walls also have the effect of reducing the effective span of the floor joists, so reduce the sectional size of the floor joist required. This will also have the effect of making the floor construction more economical in terms of timber costs.

The section size of the floor joists are determined by the

- Loading to be applied to the floor;
- Spacing between the joists;
- Span between supports (i.e. distance between sleeper walls) and
- Strength class of the timber.

By applying this information to tables given in '*Span Tables for Solid Timber Members in Floors, Ceilings and Floors (excluding trussed rafter roofs) for Dwellings*' produced by TRADA Technology Ltd a section size can be determined.

The performance requirements may be met for suspended timber ground floors by

- Covering the ground with suitable material to resist moisture and deter plant growth;
- Providing a ventilated space between the top surface of the ground covering and the timber and
- Isolating timber from moisture-carrying materials by means of damp-proof courses.

A suitable form of construction is shown in Figure 7.7 and summarised below.

- The ground surface should be covered with at least 100 mm of concrete and it should be laid on clean broken brick or similar inert hardcore not containing harmful quantities of water-soluble sulphates or other materials which might damage the concrete. Alternatively, the ground surface may be covered with at least 50 mm of concrete, as described above, or inert fine aggregate, laid on a polythene dpm as described for solid ground floors above. The joints should be sealed and the membrane should be laid on a protective bed such as sand blinding.
- Since it undesirable for water to collect on top of the ground covering material under a timber floor, the ground covering material should be laid so that *either* its top surface is not below the highest level of the ground adjoining the building *or*, where the site slopes, it might be necessary to install land drainage on the outside at the highest level of the ground adjoining the building and/or fall the ground covering material to a drainage outlet above the lowest level of the adjoining ground.
- There should be a space above the top of the concrete of at least 75 mm to any wall-plate and 150 mm to any suspended timber (or insulation where provided). This depth may

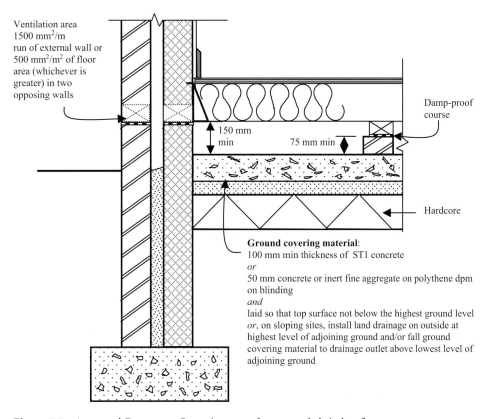

Ventilation area 1500 mm²/m run of external wall or 500 mm²/m² of floor area (whichever is greater) in two opposing walls

Damp-proof course

150 mm min

75 mm min

Hardcore

Ground covering material:
100 mm min thickness of ST1 concrete
or
50 mm concrete or inert fine aggregate on polythene dpm on blinding
and
laid so that top surface not below the highest ground level *or*, on sloping sites, install land drainage on outside at highest level of adjoining ground and/or fall ground covering material to drainage outlet above lowest level of adjoining ground

Figure 7.7   Approved Document C requirements for suspended timber floor.

need to be increased where the building is constructed on shrinkable clays in order to allow for heave.

■ There should be ventilation openings in two opposing external walls allowing free ventilation to all parts of the subfloor. An actual ventilation area equivalent to 1500 mm²/m run of external wall or 500 mm²/m² of floor area (whichever area is greater) should be provided and any ducts needed to convey ventilating air should be at least 100 mm in diameter. Ventilation openings should be fitted with grilles so as to prevent vermin entry but these grilles should not unduly resist the flow of air. It may be difficult, where there is a requirement for level access to the floor, to provide the ventilators in the position shown in Figure 7.7, since the top surface of the floor may well be nearer to the ground. The problem can usually be solved using offset (periscope) ventilators.

■ Damp-proof courses of impervious sheet materials, slates or engineering bricks bedded in cement mortar should be provided between timber members and supporting structures to prevent transmission of moisture from the ground.

■ In areas where water may be spilled (such as bathrooms, utility rooms and kitchens), boards used for flooring should be moisture resistant, irrespective of the storey in which they are located. Softwood boarding should be
o A minimum of 20 mm thick, and either
o From a durable species or
o Treated with a suitable preservative.

Chipboard is particularly susceptible to moisture damage, so where this is used as a flooring material it should be of one of the grades recommended as having improved moisture resistance

Owing to the suspect nature of timber in respect of moisture and possible rot, the introduction of more modern materials has created another alternative form of construction for ground floors, namely beam and block.

Beam and block

This form of construction is an alternative to a suspended timber floor and incorporates the use of pre-stressed concrete T beams with blocks between, as shown in Figure 7.8.

The T beams are constructed of concrete with pre-stressed steel reinforcement throughout their lengths. These beams are used in an inverted position to provide a projection for the blocks to rest on. The blocks used are standard modular blocks used for constructing walls and can be of a dense or lightweight nature.

These floors have several advantages:

- The speed of erection.
- The use of semi-skilled labour required for construction.
- A 'dry' construction as no mortars are used.
- Long spans can be achieved without the use of sleeper walls.
- A working platform can be achieved for work to proceed.

The beams are placed on top of a dpc on the inner leaf of the external wall to prevent the passage of moisture from the wall below, which will be in contact with the ground. The blocks are then placed between the beams without the use of any mortars or adhesives, so this floor is quite simple and quick to construct.

A layer of insulation will be applied to the floor when the majority of other construction activities have been completed (e.g. external walls, first floor, roof, doors and windows) and a floor finish will then be applied on top of the insulation of a sand and cement screed or particleboard.

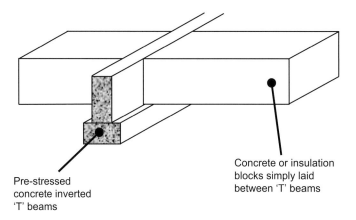

Pre-stressed
concrete inverted
'T' beams

Concrete or insulation
blocks simply laid
between 'T' beams

Figure 7.8   Beam and block floor.

If a sand and cement screed is used it will not be bonded to the structural element (i.e. the beam and block floor), so the thickness of the screed will need to be increased and reinforced with a wire mesh (Figure 7.3). This will prevent the screed from cracking and/or curling at the edges and provide sufficient strength in respect of loadings.

If a particleboard is used then a dpm is laid below the board to ensure that any moisture from below will not affect the timber.

The Building Regulations in Approved Document C give the following recommendations for suspended concrete floors:

- A damp-proof membrane should be provided if the ground below the floor has been excavated so that it is lower than outside ground level and it is not effectively drained.
- The space between the underside of the floor and the ground should be ventilated. The space should be at least 150 mm in depth (measured from the ground surface to the underside of the floor or insulation, if provided) and the ventilation recommendations should be as for suspended timber floors.
- If the building is located in an area where flooding might be a problem, it may be necessary to include a means of inspecting and clearing out the subfloor voids beneath suspended floors.

These recommendations are summarised in Figure 7.9.

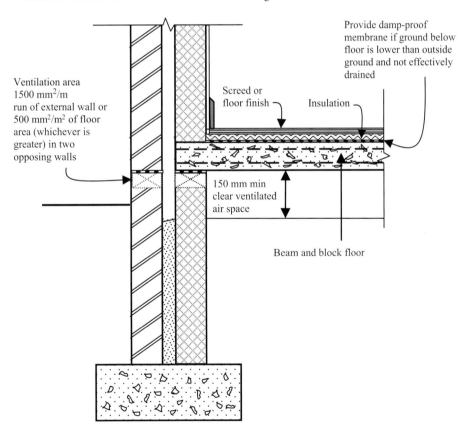

Figure 7.9 Suspended concrete ground floors.

UPPER FLOORS

Upper floors have, in the past, been quite simplistic in construction, consisting of timber joists finished with softwood boarding on the upper surface and plasterboard on the under surface. However, modern technology has influenced this simple construction.

There are several variations to the construction of an upper floor which must all satisfy the same requirements:

- Provide a level and visually acceptable wearing surface that performs the function demanded by the room in which it is situated.
- Withstand the loads placed upon it (imposed and dead).
- Provide a structure to support a ceiling below.
- Provide lateral support to the external walls of a building.
- Provide a void to locate the various services required for the dwelling.
- Provide reasonable resistance to the transmission of sound.
- In the event of a fire provide sufficient resistance to ensure the stability of the floor for a reasonable time.

There are no requirements for thermal insulation in an upper floor for a two storey domestic property.

Suspended timber upper floor

This type of floor uses solid softwood joists supported at their ends by the external walls of the building. The joists can also be supported along their lengths if there are any load-bearing partition walls within the building (i.e. an internal wall that has its own foundation and is capable of carrying loads). The joists can also be supported on joist hangers (Figure 7.10).

The section size of the floor joists are determined by the

- Loading to be applied to the floor;
- Spacing between the joists;
- Span between supports (i.e. external walls and load-bearing partition walls) and
- Strength class of the timber.

By applying this information to tables given in '*Span Tables for Solid Timber Members in Floors, Ceilings and Floors (excluding trussed rafter roofs) for Dwellings*' produced by TRADA Technology Ltd a section size can be determined. As a very rough rule of thumb, for 50 mm wide joists spaced at 400 mm centres the depth can be found by dividing the span in millimetres by 20. Therefore, to span 3.6 m (3600 mm), 50 mm wide joists spaced every 400 mm will need to be about 180 mm deep.

Joists can be liable to twisting and warping when they are spanning long distances, so intermediate strutting must be included to provide lateral stability to the floor. Different types of strutting are illustrated in Figure 7.11:

- Timber solid (also known as solid blocking)
- Timber herringbone
- Proprietary steel herringbone

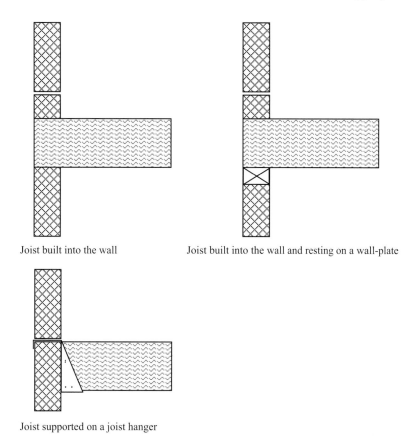

Joist built into the wall   Joist built into the wall and resting on a wall-plate

Joist supported on a joist hanger

Figure 7.10   Methods of supporting floor joists.

The spacing for the strutting is determined by the overall length of the joist and a table can also be found in '*Span Tables for Solid Timber Members in Floors, Ceilings and Floors (excluding trussed rafter roofs) for Dwellings*' produced by TRADA Technology Ltd.

Where floor joists span more than 2.5 m they should be strutted with one or more rows of

- Solid timber at least 38 mm wide and 0.75 times the joist depth or
- Herringbone strutting in 38×38 mm^2 timber except where the distance between the joists is greater than three times the joist depth.

TRADA states that proprietary herringbone strutting systems are permitted if used in accordance with manufacturer's instructions.

One row of strutting at mid-span is recommended for joist spans between 2.5 and 4.5 m. Above 4.5 m, two rows of strutting at the one third positions would be required.

TRADA adds that the outer joist should be solidly blocked to the perimeter wall at the end of each row of strutting.

Floor coverings on top of the joist can vary and traditionally would have been squared edged boards (i.e. planks). These boards evolved into tongued-and-grooved boards that would slot into each other and, therefore, would not leave gaps between them as the

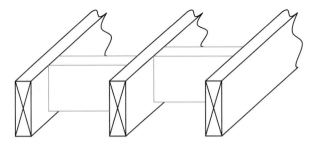

Timber solid strutting

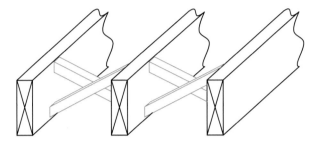

Timber herringbone strutting

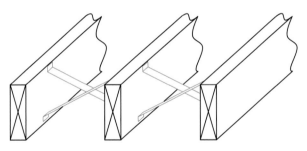

Proprietary steel herringbone strutting

Figure 7.11   Different types of strutting.

timber contracted due to thermal or moisture movement. The boards vary in widths up to about 250 mm. Common sizes are, however, around 125 mm (Figure 7.12).

Minimum softwood board thicknesses are specified in Approved Document A and vary according to the distances that they have to span between the joists. So, for joists spaced at up to 505 mm centre-to-centre the minimum board thickness is 16 mm and for spacings up to 600 mm centre-to-centre the minimum thickness is 19 mm.

The boards can be finished (e.g. varnished, oiled, waxed etc.) to provide the floor finish. If the floor finish is to be carpeted or receive some other form of finish that would leave the floorboards unseen then it is prudent these days to use a particleboard (also tongued and grooved along its edges). These boards are normally $2400 \times 1200 \, \text{mm}^2$ and have two thicknesses 18 mm (for 400 mm centred joists) and 22 mm (for 600 mm centred joists). The boards are laid at right angles to the joists and are either nailed or screwed to the joists.

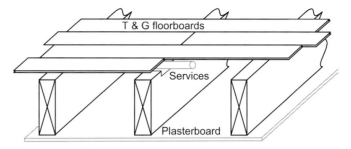

Figure 7.12 Tongue-and-groove floorboards.

The tongue-and-groove joints are also glued together with a PVA adhesive. Care must be taken to ensure that none of the screws (or nails) penetrates any of the services (hot and cold water pipes or electric cables) that may lie in the floor void between and through the joists.

The joists also provide a fixing medium in order that a ceiling can be provided for the rooms below. The most common form of ceiling is plasterboard nailed or screwed directly into the upper floor joist with the boards laid at right angles to the joist. The plasterboard can then take a variety of applied finishes. Other ceiling mediums can also be used, such as tongue-and-groove cladding, although there are restrictions in Building Regulations on the amount of combustible claddings that can be used on ceilings (outlined later). If a combustible cladding is used to cover a ceiling it would have to be underlayed by plasterboard to give the floor its fire resistance.

The floor joists provide a much needed means of lateral support to the external walls of the building. Whether they are built into the wall or whether joist hangers are used, they provide an element of rigidity and support to those walls. The walls that run parallel to the joists also require this lateral support to prevent the walls buckling or bowing throughout their heights.

This support is provided by the use of galvanised mild steel restraint straps which should be at least $30 \times 5\,\text{mm}^2$ in section. The straps have numerous pre-drilled holes to enable appropriate fixing. These 'L'-shaped straps are laid across the top of the joist with the 'L' part of the strap hooking over the internal leaf of the wall, as illustrated in Figure 7.13. A noggin

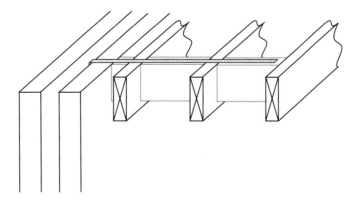

Figure 7.13 Lateral support provided by a restraint strap.

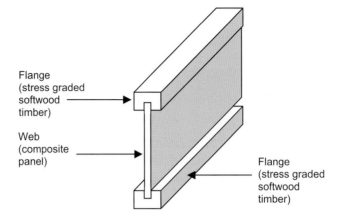

Figure 7.14   'I' joist/beam.

(the short piece of wood between the last joist and the wall) is also secured between the joists to aid fixing of the strap and to prevent the joists from moving or bowing. The restraint straps should be placed along the length of the joist span at no more than 2 m spacing.

'I' joist

An 'I' Joist is an engineered timber joist. The 'I' section is achieved by using stress graded softwood timbers to the top and bottom with a composite panel web between (Figure 7.14).

These joists have an excellent strength to weight ratio and, therefore, can allow for longer spans than traditional joists of the same section. They are easily cut and fixed using traditional woodworking tools and equipment and the webs can easily be drilled to take services.

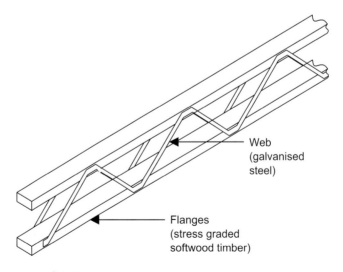

Figure 7.15   Open web joist.

Open web joist

An open web joist consists again of stress graded softwood flanges (similar to the 'I' joist). However, the flanges are separated by 'V'-shaped galvanised steel webs. This open web allows plenty of space to accommodate services without the need to drill or notch timbers and are lighter in weight than a conventional timber joist (Figure 7.15).

'I' joists and open web joists can be used for suspended ground floors as well as upper floors.

BUILDING REGULATIONS AND APPROVED DOCUMENT GUIDANCE

There are a number of issues that apply to suspended first floors in dwellings that are additional to those already discussed in this chapter. They concern

- The formation of holes and notches (Chapter 4);
- Sound insulation and
- Performance in fire.

Sound insulation

The Building Regulations in Part E require floors to have reasonable resistance to the transmission of airborne sound. For timber-joisted floors this is provided by the joist and board construction, the ceiling and the provision of an absorbent material within the construction. Approved Document E recommends the following specification:

- The timber or wood based boarding of the floor surface should have a mass per unit area of not less than 15 kg/m^2.
- The ceiling consisting of a single layer of plasterboard should have a mass of not less than 10 kg/m^2.
- There should be a layer of mineral wool laid in the cavity between the joists. The thickness should be not less than 100 mm with a density of at least 10 kg/m^2.

It should be noted that electrical cables give off heat when in use. Since thermal insulation is needed to be placed in floors and roofs, where cables are covered by thermal insulation they may overheat, increasing the risk of short circuit, or fire starting in any combustible loose fill and plastics insulation. Cables should be fixed to the structure above the insulation so that they can dissipate heat. The circuits most likely to be affected are radial circuits serving cookers, immersion heaters, shower units and socket outlets.

Performance in fire – surface spread of flame

The Building Regulations in Regulation B2 control the amount of combustible material that can be placed on the surfaces of walls and ceilings in dwellings.

Table 1 from Section 3 of Volume 1 of Approved Document B gives the recommended flame spread classifications for the surfaces of walls and ceilings in any room or circulation space.

Different standards are set for 'small rooms', which are totally enclosed rooms with floor area of not more than $4\,m^2$ and for other rooms and circulation spaces. Small rooms are allowed to have surface linings of not lower than Class 3 and other rooms are only permitted to have linings that are not lower than Class 1 (although a slight relaxation of this is permitted to allow a surface lining of no lower than Class 3 in these other rooms provided that it does not cover an area more than half the floor area of the room). These classes relate to the speed at which a fire would spread across the surface of the lining and would, therefore, have the ability to spread a fire rapidly throughout a room.

Common ceiling materials that come within Class 1 include plasterboard and mineral fibre tiles. Common materials that come within Class 3 include timber or plywood with a density of more than $400\,kg/m^3$ whether it is painted or not. If it is desired to fully line a ceiling in a room which is larger than $4\,m^2$ in area then proprietary clear treatments are available; when applied to the surface of the wood these treatments improve its surface spread of flame class to Class 1.

Performance in fire – fire resistance

If the structural elements of a building can be satisfactorily protected against the effects of fire for a reasonable period, it will be possible for the occupants to be evacuated safely. Also, the spread of fire throughout the building will be kept to a minimum. The risk to firefighters (who may have to search for or rescue people who are trapped) will be reduced and there will be less risk to people in the vicinity of the building from falling debris or as a result of an impact with an adjacent building from the collapsing structure.

One way to measure the standard of protection to be provided is by reference to the fire resistance of the elements under consideration. Fire resistance is measured in the number of minutes that a sample of the element (wall, floor etc.) performs in a standard test. The test relates to the ability of the element to resist

- A fire without collapse (load-bearing capacity);
- Fire penetration (integrity) and
- Excessive heat penetration so that fire is not spread by radiation or conduction (insulation).

For floors in dwellings of not more than two storeys the floor is required to have what is termed a 'modified half hour' of fire resistance. For a timber-joisted floor this would result in the following specification:

- Load-bearing capacity – 30 min, integrity 15 min, insulation 15 min.

Specification 1

- Any structurally suitable flooring.
- Aloor joists at least 37 mm wide.
- Ceiling:
  o 12.5 mm plasterboard with joints taped and filled and backed by timber or
  o 9.5 mm plasterboard with 10 mm lightweight gypsum plaster finish.

Specification 2

- At least 15 mm tongue-and-groove boarding or sheets of plywood or wood chipboard.
- Floor joists at least 37 mm wide.
- Ceiling:
    - 12.5 mm plasterboard with joints taped and filled or
    - 9.5 mm plasterboard with at least 5 mm neat gypsum plaster finish.

FURTHER INFORMATION

References used in this chapter include the following:

- *Span Tables for Solid Timber Members in Floors, Ceilings and Floors (excluding trussed rafter roofs) for Dwellings*, produced by TRADA Technology Ltd. ISBN: 1 900510 42 1
- Building Regulations Approved Document A – Structure
- Building Regulations Approved Document B – Fire Safety
- Building Regulations Approved Document E – Resistance to the Passage of Sound

Further information can be obtained from

www.trada.co.uk
http://www.milbank.co.uk/
http://www.rackhamhousefloors.co.uk/
http://www.planningportal.gov.uk/

8 External walls

Questions addressed in this chapter:

How can I construct an external wall?
Are there any new materials and methods in use for these constructions?
What Building Regulations apply to these construction areas?

INTRODUCTION

The external walls of a building must satisfy a number of functional requirements, including the following:

- Ability to carry loads and distribute them safely to the foundations
- Provision of an acceptable level of thermal insulation
- Provision of an acceptable level of sound insulation
- Ability to keep out the elements of the weather
- Ability to resist the adverse affects of rising ground water
- Standard of durability suitable for the life of the building
- Structural stability
- Provision of an appropriate level of fire resistance
- Appropriate aesthetic appearance (especially in conservation areas and when dealing with listed buildings).

Most of these requirements are covered by Building Regulations (see the section at the end of this chapter) and other legislation (Chapter 2). However, it is interesting to note that while a degree of sound insulation is desirable in external walls so as to provide a satisfactory, and reasonably quiet, internal environment, current building regulations do not cover sound penetrating from outside sources. This tends to be dealt with by other legislation, especially where new dwellings are to be erected close to existing roads or airports.

External walls have been constructed using many types of materials and many methods over the years. Therefore, it is not uncommon for the main building to be constructed differently to any later extensions or adaptations.

With brick being manufactured by the Egyptians thousands of years ago and stone and timber being natural building products, these have been common walling materials. However, different types of block and man-made stones, among others, have provided a greater choice of walling materials. Also, what may be a common walling material in one country may not be in another, since local traditions in construction tend to reflect the presence of local, naturally occurring materials. For example, in England there is a tendency to build walls using clay bricks, since there is a preponderance of clay, especially

Extending and Improving Your Home: An Introduction, First Edition. M.J. Billington and C. Gibbs.
© 2012 M. J. Billington and C. Gibbs. Published 2012 by Blackwell Publishing Ltd.

in the South of the country. Houses constructed using clay bricks are comparatively rare in the Republic of Ireland due to the relative scarcity of clay in that country. The types of external walls include solid walls, cavity walls and timber frame.

Masonry walls

The term 'masonry' is now used to describe any form of construction using bricks, blocks or stone bonded together with lime or cement mortar. The term derives from the craftsman (mason) who would have carried out the work. Originally it applied only to stonemasons. Nowadays, people who lay bricks are, unsurprisingly, referred to as bricklayers and not masons.

SOLID WALLS

Solid walls encompass many different materials and are commonly linked with earlier forms of construction. However, solid walls can still be used adequately in modern-day construction. Early forms of solid walls were commonly constructed of brickwork or stonework depending on the locality and availability of materials.

Solid brick walls

Solid brick walls would often be constructed to a thickness of 1–1½ bricks. One brick thick means that the wall has a thickness equivalent to the length of a brick. Traditionally, bricks were standardised at a length of 9 in. (225 mm), a width of 4½ in. (112.5 mm) and a depth of 3 in. (75 mm). Therefore, a one brick wall was 9 in. thick and a 1½ brick wall was 13½ in. thick. The long side of a brick is also referred to as a stretcher and the short side (i.e. the width of the brick) as a header, especially when talking about brick bonds. Although there are many named bonds, generally only two bonds are most common, these being the quarter bond and the half bond. The terms quarter bond and half bond refer to the amount of overlap one brick would have over the brick below in the bonding arrangement. Therefore, the bricks in a half bond wall would overlap a distance equivalent to that of half a brick (e.g. stretcher bond, Figure 8.6) and, of course, bricks in a quarter bond wall would overlap each other by the equivalent distance of quarter of a brick (e.g. English bond, Figure 8.3).

As there are two headers to one stretcher, if these were laid directly above each other, there would be a straight joint, which would reduce the strength of the brickwork. To prevent this from occurring, a Queen closer is placed next to the quoin header. The 'quoin header' is the header brick that forms the quoin (or corner) of the wall. A 'queen closer' is a brick cut lengthways to form a quarter brick width. This ensures that a quarter bond is achieved (Figure 8.1).

A large variety of different bonds were developed for constructing brick walls, especially during Victorian times. However, Flemish bond was most commonly used for domestic housing, as this afforded adequate strength with aesthetic appeal (as the headers were often picked out in a different coloured brick to improve the decorative appearance) (Figure 8.2).

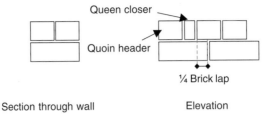

Queen closer

Quoin header

¼ Brick lap

Section through wall   Elevation

Figure 8.1   A quarter bond.

Flemish bond consists of alternating headers and stretchers in the same course (layer of brickwork). Flemish bond is a quarter bond, which means that the bricks overlap the bricks below by a quarter of a brick. The header in the course would be located centrally over the stretcher in the course below.

English bond is another quarter bond and is stronger than Flemish bond owing to there being no internal straight joints. This superior strength was not required for domestic housing and this bond was generally reserved for more structural works, such as bridges and retaining walls. This bond consists of alternating courses (or layers) of headers and stretchers (Figure 8.3).

There are variations to both English and Flemish bonds that are called garden wall bonds. Both of these bonds use more stretchers in the construction of the wall. By doing this and reducing the number of headers in the wall, an improved finish can be maintained to both sides of the wall, which could not be achieved with either English or Flemish bond. This explains why they are called garden wall bonds, as these bonds would be used for walls that separate gardens and, therefore, both properties could have a fairface (or good finish) to the wall (Figure 8.4).

Although external solid brickwork provides a wall that is structurally stable, it does not provide a good level of thermal insulation (thermal performance) and can be poor at preventing the passage of moisture to the internal environment, especially once the mortar courses between the bricks begin to deteriorate with age and lack of adequate maintenance. It is for this reason that solid brick walls are not used today and why existing solid brick-walled houses are often rendered to help resist the weather and may have thermal insulation applied to either the internal or the external face (further information on improving the thermal efficiency of the home is given in Chapter 13).

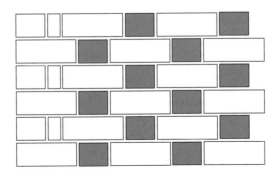

Figure 8.2   Flemish bond.

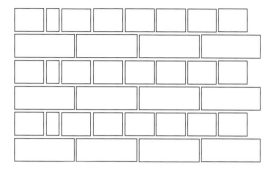

Figure 8.3 English bond.

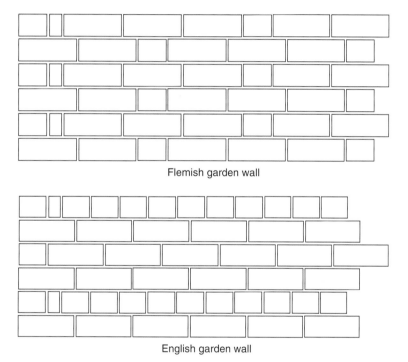

Flemish garden wall

English garden wall

Figure 8.4 Garden wall bonds.

Solid stone walling

Stone was a common walling material using types such as sandstone, limestone, granite and even flint. Because stone is expensive to transport, it is found that certain areas of the country have traditional walling types based on their own locally produced stones, and after the dissolution of the monasteries in the 1500s, many of the stones from the earlier Norman monasteries were used by local people to build their own cottages. A fine example of recycling materials! The softer stones, such as sandstone and limestone, were much easier to work with and therefore could be 'dressed' to shape and have square edges. Where this type of dressed stone exists, it is referred to as dressed stonework. The finest examples of this in towns such as Bath and Cheltenham were built in what is referred to as 'ashlar' stonework. However, the more dense stones, such as granite and flint, were often left in their irregular shape and this would be called random stonework. In parts of the country, such as the Lake District and in the Channel Islands where granite is prevalent, the granite was often squared off and laid in courses while flint, which commonly occurs in Sussex and Norfolk, was often used as a facing material backed up by clay brickwork. Owing to the shape and nature of stone, external walls were constructed in two parts with a rubble (and, not uncommonly, clay) infill between the internal and external walling. To prevent the two halves of these walls separating and therefore bowing, larger stones, often called

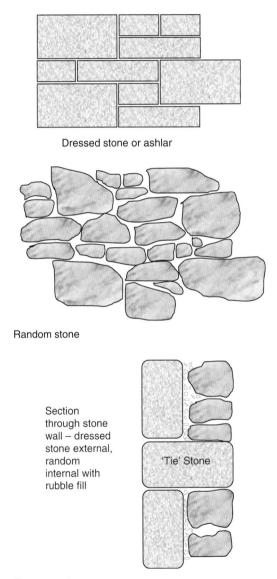

Dressed stone or ashlar

Random stone

Section through stone wall – dressed stone external, random internal with rubble fill

'Tie' Stone

Figure 8.5　Stone wall construction.

'through' stones or 'tie' stones, would be used across the thickness of the two walls to tie them together (Figure 8.5).

As explained, whichever walling material is used, the external walling may be dressed or random stonework. However, the internal walling would normally be random stonework, as this was considerably cheaper to construct, used up all the left over pieces of stone and would often be covered over with a lime mortar render. It is common these days, however, to expose this stonework and re-point it to give a pleasing decorative appearance. Where walls are constructed using lime mortar (as is generally the case for older construction for both stonework and brickwork), it is preferable to continue to use a lime mortar for any work such as pointing or rendering, as lime mortars not only allow better evaporation of moisture but also flexibility in the construction, which reduces the amount of cracking that could occur owing to the flexible nature of the wall itself.

Cement mortars tend to be much stronger and, therefore, more brittle; so should any movement occur within the structure, the pointing or render could crack and fall away from the wall.

As with solid brick walls, stone walls offer poor thermal insulation and can allow moisture penetration to the interior.

CAVITY WALLS

As technology advanced and alternative forms of construction were developed, cavity walls were introduced. Cavity walls were commonplace from the late 1930s, although they are not unknown in houses before that time. As the name suggests, this type of wall includes a cavity that is an air space that separates two individual walls (one external and one internal), which are called leaves (Figure 8.6). Both leaves (or skins) of brickwork were constructed of half brick walls in Stretcher bond. The external leaf and internal leaf were tied together using steel wall ties. These wall ties were made of mild steel coated with zinc (a process known as galvanising) to protect the ties from rusting. Galvanised wall ties are reasonably resistant to rusting, since if the zinc coating is thick enough, it has the ability to repair itself should it become damaged. In the 1970s, a change in the code for the manufacture of galvanised wall ties led to a reduction in the allowable thickness of zinc. Some years later there were some spectacular building failures and in one instance the entire outer leaf of a dwelling collapsed due to massive amount of degradation of the wall ties. This was also linked to the use of aggressive colourants in the mortar which exacerbated the problem of the reduced standard of galvanising. This occurrence highlighted the problem and a great many houses were investigated and also found to be defective. This led to a large number of houses being treated for wall tie failure using a number of different techniques specifically developed for this problem. Modern wall ties can still be galvanised but it is also possible to use stainless steel or plastic wall ties, which would be preferable as these will not corrode (Figure 8.7).

Although the inclusion of the cavity air space between the two leaves in a cavity wall did not offer a large improvement in thermal insulation, it did offer a much improved resistance to the passage of moisture, as moisture passing through the external leaf did not have a direct passage through the inner leaf except via the wall ties. This is why wall ties will

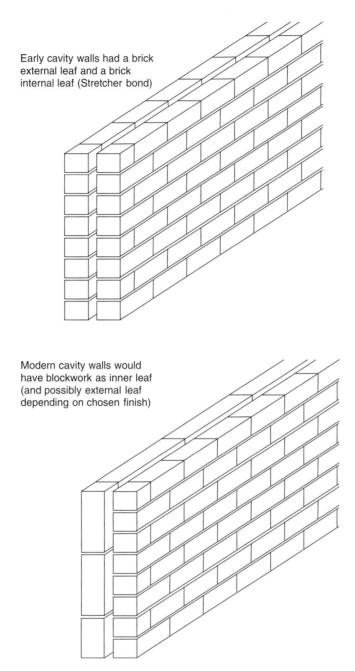

Early cavity walls had a brick external leaf and a brick internal leaf (Stretcher bond)

Modern cavity walls would have blockwork as inner leaf (and possibly external leaf depending on chosen finish)

Figure 8.6 Cavity wall construction.

always have a twist or drip on them, which should be placed at the centre of the cavity so that should any water pass along the tie, it will drip off and fall down the cavity and go back to the outside. In addition, it is common practice when constructing cavity walls to ensure that the wall ties slope slightly towards the external leaf to prevent water passing back along the ties to the inner leaf (Figure 8.7).

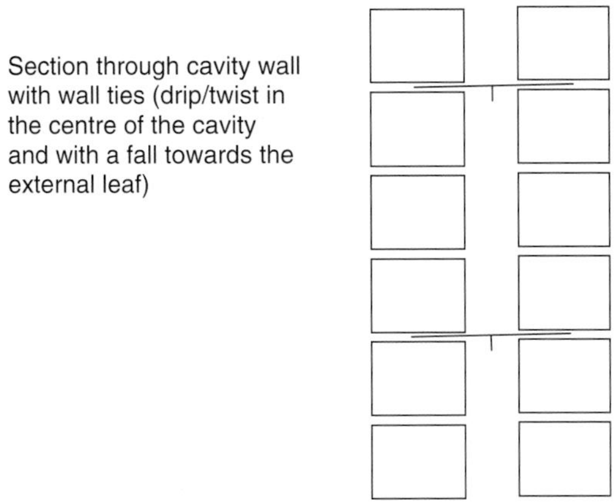

Section through cavity wall
with wall ties (drip/twist in
the centre of the cavity
and with a fall towards the
external leaf)

Figure 8.7 Wall ties.

As modern standards have improved over time, the cavity is commonly filled, either partially or totally, with insulation (Figure 8.8). This now provides a wall with improved thermal performance and improved resistance to the passage of moisture.

The problem with filling (or partially filling) the cavity in the wall with insulation is that it rather defeats the object of having a cavity in the first place, since this was originally introduced to prevent the passage of moisture through the wall. Therefore, it is extremely important to follow good building practice when constructing insulated cavity walls. Mortar must not be allowed to collect on the surface of

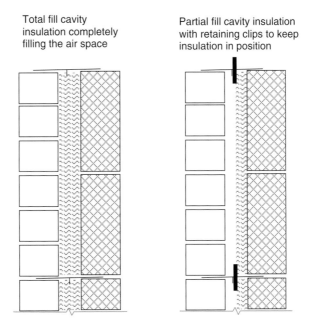

Total fill cavity
insulation completely
filling the air space

Partial fill cavity insulation
with retaining clips to keep
insulation in position

Figure 8.8 Cavity wall insulation.

the insulation batts as the wall is built as it will provide a ready path for the passage of moisture. Also, when partial fill is used, the insulation-retaining clips must be properly fixed so that the batts are not allowed to tip forward again creating a path for moisture.

Inserting the insulation in the cavity does not greatly affect the thickness of the wall. However, insulation could also be placed on the internal or external face of the cavity wall. These locations, however, are generally reserved for refurbishment work where access to the cavity is limited. The cavity can also be filled with insulation after construction by injection. Holes are drilled into the wall and mineral wool fibres or sticky polystyrene beads are blown into the cavity to fill the air space.

The techniques of external insulation and injected cavity wall insulation are normally undertaken by specialist contractors and it is important that reputable contractors are used for this who will offer guarantees for their work.

While cavity walls are regarded as standard construction these days, some block manufacturers are promoting solid wall construction using aerated concrete blocks. These blocks offer a high thermal performance and a large increase in the speed of construction. They do not, however, provide any decorative finish, so an applied finish such as a render must be used.

FRAMED WALL CONSTRUCTION

Timber frame

Timber frame wall panels are commonly made up of softwood vertical studs (vertical pieces of timber) and horizontal rails with a wood-based panel sheathing (lining board) and a plasterboard internal lining. The studs carry vertical loads through the structure and transfer them to the foundations. The sheathing provides resistance to lateral wind loads (known as racking resistance). Thermal insulation is usually incorporated in the spaces between the studs of external walls and protective membrane materials may also be required, depending on the design of the wall.

For most external walls, a breather membrane on the external face of the panels protects the panels during construction and provides a second line of defence against any wind-driven rain that may penetrate the completed external cladding. A vapour control layer in the form of polythene sheet or plasterboard with an integral vapour control layer is normally required on the 'warm' side of the insulation, behind the plasterboard lining, to limit the amount of water vapour entering the wall panel. The advantages of timber frame wall construction are as follows:

- Speed of erection
- The ability to manufacture the panels in a factory under ideal conditions
- Accuracy of construction
- The ability to provide high levels of insulation without unduly increasing the thickness of the wall
- Dry construction

Typical panel details are shown in Figure 8.9 with different external finishes.

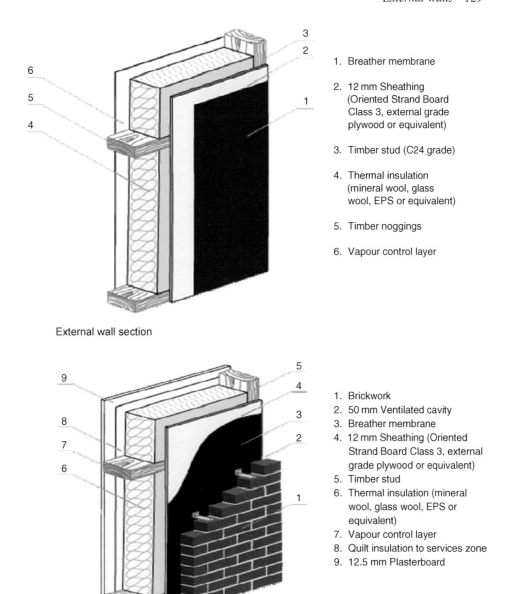

External wall section

1. Breather membrane

2. 12 mm Sheathing (Oriented Strand Board Class 3, external grade plywood or equivalent)

3. Timber stud (C24 grade)

4. Thermal insulation (mineral wool, glass wool, EPS or equivalent)

5. Timber noggings

6. Vapour control layer

1. Brickwork
2. 50 mm Ventilated cavity
3. Breather membrane
4. 12 mm Sheathing (Oriented Strand Board Class 3, external grade plywood or equivalent)
5. Timber stud
6. Thermal insulation (mineral wool, glass wool, EPS or equivalent)
7. Vapour control layer
8. Quilt insulation to services zone
9. 12.5 mm Plasterboard

Brick–Clad finish with 50 mm Cavity

Figure 8.9 Typical panel details for timber frame construction.

Structural insulated panel wall systems

Structural insulated panels (SIPS) are an advanced method of construction, offering superior insulation, structural strength and airtightness over traditional construction methods or systems. SIPS can be used in floors, walls and roofs for most types

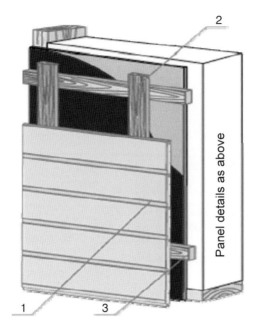

Panel details as above

1. 18 mm Horizontal timber cladding

2. 25 × 50 mm Vertical batten max 600 mm centres

3. 25 × 50 mm Horizontal batten max 600 mm centres

Horizontal timber board finish

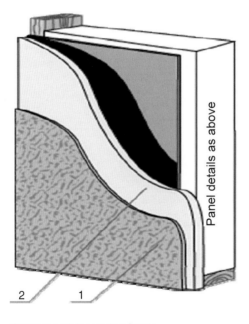

Panel details as above

1. 6–8 mm Render

2. 50 mm/100 mm Thermal insulation – rigid batts of rockwool or glass fibre or EPS

External wall insulation finish

Figure 8.9   (*Continued*)

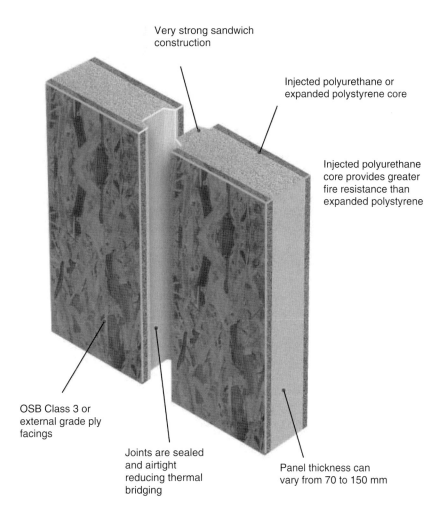

Very strong sandwich
construction

Injected polyurethane or
expanded polystyrene core

Injected polyurethane
core provides greater
fire resistance than
expanded polystyrene

OSB Class 3 or
external grade ply
facings

Joints are sealed
and airtight
reducing thermal
bridging

Panel thickness can
vary from 70 to 150 mm

Figure 8.10 Structural insulated panels.

of building, as well as for extensions to dwellings. Typical panels are shown in
Figure 8.10.

The panels are structural elements usually consisting of internal and external skins of
Class 3 oriented strandboard (OSB3) with an insulation core of closed-cell polyurethane
(PUR) or expanded polystyrene (EPS). The method of manufacture differs according to
the type of insulation used. For injected polyurethane, the OSB sheets are held apart in a jig
and the polyurethane is injected into the space between them. As the chemical reaction
takes place, the polyurethane foams and expands to fill the space, at the same time bonding
to both sheets. This forms a very strong mechanical bond and the sheets and core act as one
unit. With EPS, the preformed sheets of insulation are glued to each of the OSB sheets.
This is a simpler method of manufacture and does not need complicated and expensive

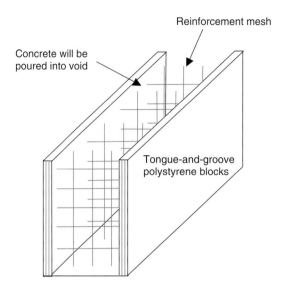

Reinforcement mesh

Concrete will be
poured into void

Tongue-and-groove
polystyrene blocks

Figure 8.11   An Insulated Concrete Form (ICF) unit.

machinery to form the panels. SIPS panels made using EPS are not as inherently strong as injected polyurethane panels and they are not as fire resistant.

The panels are for use above the damp-proof course in domestic applications (including rooms-in-roof) as the load-bearing inner leaf of an external cavity wall. In effect, they replace the structural studding, insulation and vapour control layer of timber frame construction, since the OSB3 used in their fabrication can act as a vapour control layer if the joints between the boards are sealed with vapour-proof tape.

Insulated Concrete Forms

Insulated Concrete Forms (ICFs) are becoming more popular with builders and designers as they can offer a quick erection sequence using light manageable units. These units (approximately 1200 mm long, 400 mm high and 300 mm wide) comprise two 70 mm thick moulded polystyrene (EPS) internal and external forms with a separating web. These units are filled with reinforced concrete to give the necessary structural performance. This construction is more common for a complete new build rather than an extension or improvement. ICFs can also be used for basement construction when a waterproof concrete (ordinary concrete with a waterproofing additive) would be used (Figure 8.11).

BUILDING REGULATIONS AND APPROVED DOCUMENT GUIDANCE

At the start of this chapter, the functional and performance requirements of external walls were referred to. From the items listed, building regulations apply to the following:

- Ability to carry loads and distribute them safely to the foundation and provide an adequate level of structural stability (Part A – Structure)

- Provision of an acceptable level of thermal insulation (Part L – Conservation of Fuel and Power);
- Ability to keep out the elements of the weather and resist the adverse affects of rising groundwater (Part C – Site Preparation and Resistance to Contaminants and Moisture)
- Standard of durability suitable for the life of the building (c Regulation 7)
- Provision of an appropriate level of fire resistance (Part B – Fire Safety)

Structural stability

Part A has already been discussed in Chapter 6 where the requirements for foundations were looked at. For walls, most traditional brick and block constructions are comparatively easy to deal with. The principal factors governing their strength and stability are the thickness of the wall relative to its height and length and the strength of the units (bricks, blocks or stone) that make up the wall.

The minimum thicknesses required depend upon the wall height and length, and the rules applying to walls of bricks or blocks are set out in Table 8.1 (Table 3 of Approved Document A) and illustrated in Figure 8.12.

For the strength of the blocks or bricks, for normal extensions up to two storeys high, it is sufficient to use blocks with a minimum compressive strength of $2.8\,\text{N/mm}^2$ (Newtons per square millimetre) and bricks with a minimum compressive strength of $5\,\text{N/mm}^2$. For exceptional conditions (below ground walls greater than 1 m high or floor to ceiling heights greater than 2.7 m), it would be necessary for the minimum strengths to be increased to $7\,\text{N/mm}^2$ for both bricks and blocks. For designs outside these dimensions, a structural engineer should be employed to calculate the loadings, stresses and so on.

As a general rule, the thickness of any storey of a brick or block wall should not be less than 1/16th of the height of that storey. However, walls of uncoursed stone, flints, clunches of bricks (bricks which fused together when they were taken from the kiln) or other burnt or vitrified material should have a thickness of at least $1\frac{1}{3}$ times the thickness required of brick or block walls. Irrespective of the materials used in

Table 8.1 Approved Document A, Table 3: Minimum thickness of certain external walls, compartment walls and separating walls.

(1) Height of wall	(2) Length of wall	(3) Minimum thickness of wall
Not exceeding 3.5 m	Not exceeding 12 m	190 mm for the whole of its height
Exceeding 3.5 m but not exceeding 9 m	Not exceeding 9 m	190 mm for the whole of its height
	Exceeding 9 m	290 mm from the base for the height of one storey and 190 mm for the rest of its height
Exceeding 9 m but not exceeding 12 m	Not exceeding 9 m	290 mm from the base for the height of one storey and 190 mm for the rest of its height
	Exceeding 9 m but not exceeding 12 m	290 mm from the base for the height of two storeys and 190 mm for the rest of its height

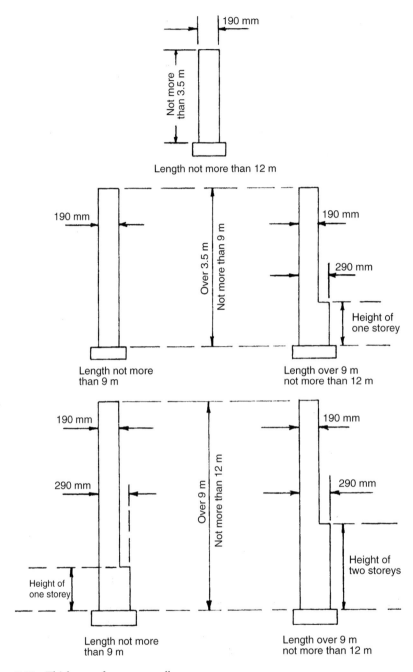

190 mm

Not more than 3.5 m

Length not more than 12 m

190 mm

Over 3.5 m
Not more than 9 m

190 mm

290 mm

Height of one storey

Length not more than 9 m

Length over 9 m not more than 12 m

190 mm

290 mm

Over 9 m
Not more than 12 m

190 mm

290 mm

Height of two storeys

Height of one storey

Length not more than 9 m

Length over 9 m not more than 12 m

Figure 8.12 Thickness of masonry walls.

construction, no part of a wall should be thinner than any other part of the wall that it supports.

For cavity walls, the two leaves should be properly tied together with wall ties in compliance with Table 5 of Approved Document A (an extract is given in Table 8.2).

Table 8.2 Extract from Table 5 of Approved Document A – Cavity Wall Ties.

Normal cavity width (mm)	Permissible type of tie	
	Tie length (mm)	Tie shape in accordance with BS 1243
60–75	200	Butterfly, double triangle or vertical twist
76–95	225	Double triangle or vertical twist
91–100	225	Double triangle or vertical twist
101–125	250	Vertical twist
126–150	275	Vertical twist
151–175	300	Vertical twist
176–300	See note*	Vertical twist style

*The depth that the tie is embedded should not be less than 50 mm in both leaves. For cavity widths wider than 180 mm, the minimum length of the tie should be the structural cavity width plus 150 mm.

Ties should be placed at centres 900 mm horizontally and 450 mm vertically (i.e. 2.5 ties/m^2), and at any opening, movement joint or roof verge at least one tie should be provided for each 300 mm of height within 225 mm of the opening. The cavity should be at least 50 mm wide and each leaf should be at least 90 mm thick at any level. The sum of the thicknesses of the two leaves, plus 10 mm, should not be less than the thickness required for a solid wall of the same height and length by Table 3 of Approved Document A (Table 8.1). Figure 8.13 and Table 8.2 illustrate this.

Special rules apply to the construction of small detached buildings such as garages, sheds and annexes to dwellings. These are illustrated in Figure 8.14.

Design for thermal efficiency

Part L of the Building Regulations covers the construction of extensions to dwellings and gives guidance on the thermal performance expected in Approved Document L1B, Conservation of Fuel and Power in Existing Dwellings. For simple designs without excessive window and door areas, the guidance is quite straightforward. However, where the area of windows and doors exceeds 25% of the floor area of the extension, it will be necessary to carry out a thermal calculation and it is wise to use the services of an expert to do this, as it involves complicated calculations and an understanding of the thermal properties of the materials used.

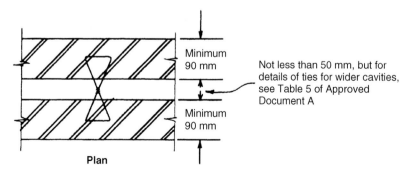

Figure 8.13 Cavity wall tie sizes related to cavity widths in masonry cavity walls.

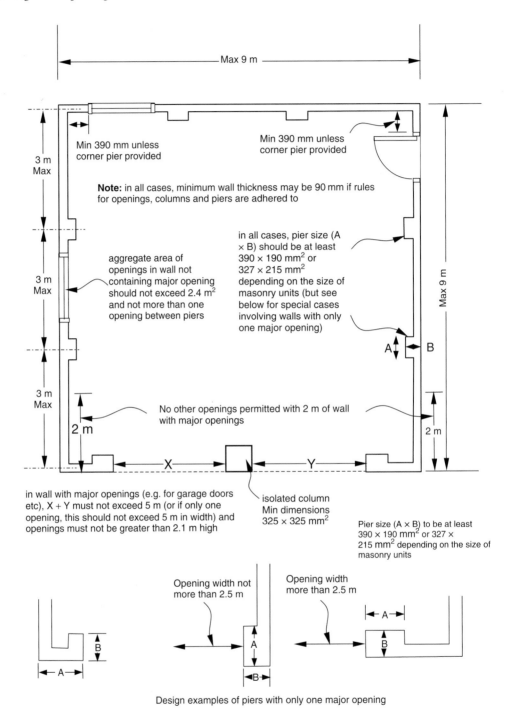

Figure 8.14   Small buildings and annexes – design of openings, piers and columns.

Table 8.3 Extract from Table 2 of Approved Document L1B – Standards for New Thermal Elements.

Element	U-value Standard (W/m²K)
Wall	0.28
Pitched roof – insulation at ceiling level	0.16
Pitched roof – insulation at rafter level	0.18
Flat roof or roof with integral insulation	0.18
Floors	0.22
Window, roof window or rooflight	1.6*
Doors	1.8

* or Window Energy Rating (WER) of Band C or better.

Firstly, the thermal elements of the extension (roof, walls, ground floor, external windows and doors) must achieve certain thermal values termed U-values (a U-value is called a thermal transmittance coefficient and the larger the number, the worse the standard of performance). The U-values that must not be exceeded are shown in Table 2 from Approved Document L1B (an extract from this is shown in Table 8.3).

Secondly, to avoid having to carry out thermal calculations on the extension, the area of new windows and doors should be restricted to no more than 25% of the newly created extension floor area. This area can be increased if, as a result of the extension, windows and doors that were in the original house are covered by the extension. So the areas of such windows and doors can be added to the 25% floor area allowance without incurring the need for a thermal calculation.

Weather and damp resistance

Part C of the Building Regulations requires the floors, walls and roof of a building to adequately protect it and the people who use it from harmful effects caused by

- Moisture from the ground;
- Precipitation and wind-driven spray;
- Surface and interstitial condensation and
- Water spilt from or associated with sanitary fittings and fixed appliances.

For walls, we are concerned with resisting rising groundwater and penetration from rain and snow, which can cause dampness in the walls and damage to plaster and decorative finishes. Approved Document C has already been discussed with reference to ground floors in Chapter 7, but it also covers external walls.

The term *wall* includes piers, columns and parapets and may include chimneys if they are attached to the building. Windows, doors and other openings are not included, but the joint between the wall and the opening is included.

Walls should be constructed so that

- The passage of moisture from the ground to the inside of the building is resisted;
- They will not be adversely affected by moisture from the ground and
- They will not transmit moisture from the ground to another part of the building that might be damaged.

The requirements mentioned above can be met for internal and external walls by providing a damp-proof course of suitable materials in the required position. Figure 8.15 illustrates the main provisions, which are summarised below.

The damp-proof course may be of any material that will prevent moisture movement. This would include bituminous sheet materials, engineering bricks or slates laid in cement mortar, polyethylene or pitch polymer materials.

The damp-proof course and any damp-proof membrane in the floor should be continuous.

Unless an external wall is suitably protected by another part of the building, the damp-proof course should be at least 150 mm above outside ground level.

Where a damp-proof course is inserted in an external cavity wall, the cavity should extend at least 225 mm below the level of the lowest damp-proof course. Alternatively, precipitation can be prevented from reaching the inner leaf of the wall by the use of a damp-proof tray. This may be particularly useful where a cavity wall is built

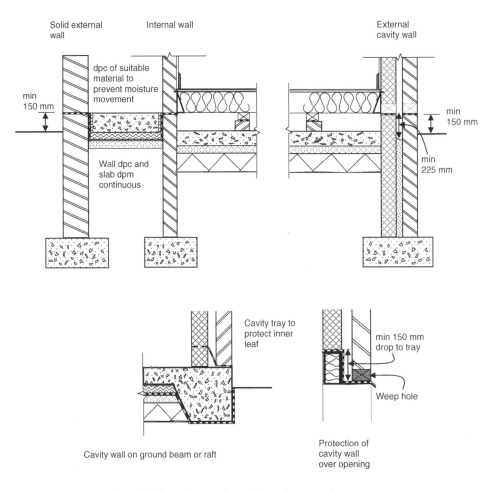

Figure 8.15   Protection of walls against moisture from the ground.

directly off a raft foundation, ground beam or similar supporting structure, and it is impractical to continue the cavity down 225 mm. Where a cavity tray is inserted, weep holes should be provided every 900 mm in the outer leaf to allow moisture collecting on the tray to pass out of the wall. In some circumstances, such as above a window or door opening, the cavity tray will not extend the full length of the exposed wall. Here, stop ends should be provided to the tray and at least two weep holes should be provided.

In addition to resisting ground moisture, external walls should

- Resist the passage of precipitation to the inside of the building;
- Not transmit moisture due to precipitation to other components of the building that might be damaged;
- Be designed and constructed so as not to allow interstitial condensation to adversely affect their structural and thermal performance and
- Not promote surface condensation and mould growth under reasonable occupancy conditions.

There are a number of forms of wall construction which will satisfy the above requirements:

- A solid wall of sufficient thickness holds moisture during bad weather until it can be released in the next dry spell.
- An impervious or weatherproof cladding prevents moisture from penetrating the outside face of the wall.
- The outside leaf of a cavity wall holds moisture in a similar manner to that of a solid wall, the cavity preventing any penetration to the inside leaf.

These principles are illustrated in Figure 8.16.

The thickness of a solid external wall will depend on the type of brick or block used and the severity of exposure to wind-driven rain. In conditions of *very* severe exposure, it may be necessary to use an external cladding. However, in conditions of severe exposure, a solid wall may be constructed as shown in Figure 8.17.

The following points should also be considered:

- For brickwork or stonework, the wall should be at least 328 mm thick.
- For dense aggregate blockwork, the wall should be at least 250 mm thick.
- For lightweight aggregate or aerated autoclaved concrete, the wall should be at least 215 mm thick.
- The brickwork or blockwork should be rendered or given an equivalent form of protection.
- Rendering should have a scraped or textured finish and be at least 20 mm thick in two coats; this permits easier evaporation of moisture from the wall.
- The bricks or blocks and mortar should be matched for strength to prevent cracking of joints or bricks and joints should be raked out to a depth of at least 10 mm to provide a key for the render.
- The render mix should not be too strong or else cracking may occur; a mix of 1:1:6 cement:lime:well-grade sharp sand is recommended for all walls except those constructed of dense concrete blocks where 1:½:4 should prove satisfactory.

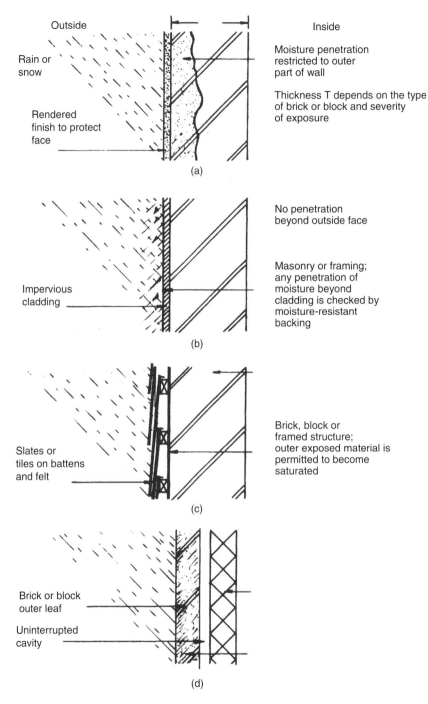

Outside

Rain or
snow

Rendered
finish to protect
face

(a)

Inside

Moisture penetration
restricted to outer
part of wall

Thickness T depends on the type
of brick or block and severity
of exposure

Impervious
cladding

(b)

No penetration
beyond outside face

Masonry or framing;
any penetration of
moisture beyond
cladding is checked by
moisture-resistant
backing

Slates or
tiles on battens
and felt

(c)

Brick, block or
framed structure;
outer exposed material is
permitted to become
saturated

Brick or block
outer leaf

Uninterrupted
cavity

(d)

Figure 8.16   Principles of weather resistance of external walls: (a) solid external wall; (b) impervious cladding; (c) weather-resistant cladding; (d) cavity wall.

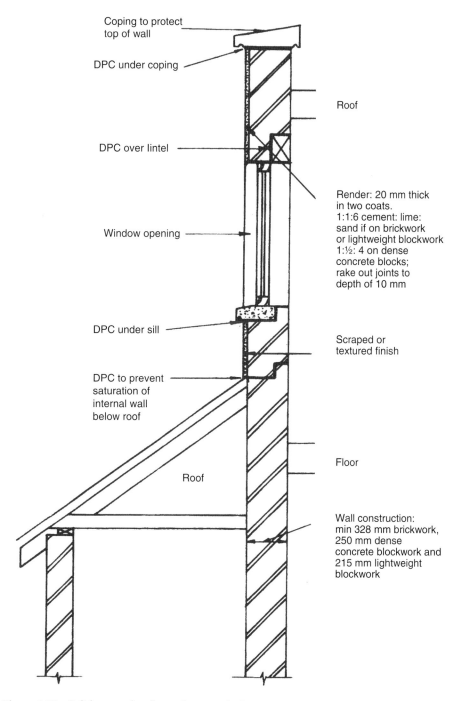

Coping to protect
top of wall

DPC under coping

Roof

DPC over lintel

Render: 20 mm thick
in two coats.
1:1:6 cement: lime:
sand if on brickwork
or lightweight blockwork
1:½: 4 on dense
concrete blocks;
rake out joints to
depth of 10 mm

Window opening

DPC under sill

Scraped or
textured finish

DPC to prevent
saturation of
internal wall
below roof

Roof

Floor

Wall construction:
min 328 mm brickwork,
250 mm dense
concrete blockwork and
215 mm lightweight
blockwork

Figure 8.17 Solid external walls: moisture exclusion.

It is, of course, possible to obtain a wide range of premixed and proprietary mortars and renders. These should be used in accordance with the manufacturer's instructions. The following guidance should also be followed:

- Where the top of a wall is unprotected by the building structure, it should be protected to resist moisture from rain or snow. Unless the protection and joints form a complete barrier to moisture, a damp-proof course should also be provided.
- Damp-proof courses, trays and closers should be provided to direct moisture towards the outside face of the wall in the positions shown in Figure 8.17.
- Insulation to solid external walls may be provided on the inside or outside of the wall. Externally placed insulation should be protected unless it is able to offer resistance to moisture ingress so that the wall may remain reasonably dry (and the insulation value may not be reduced). Internal insulation should be separated from the wall construction by a cavity to give a break in the path for moisture. (Some examples of external wall insulation are given in Figure 8.18.)

For external cavity walls, to meet the performance requirements, the wall should consist of an internal leaf which is separated from the external leaf by

- A drained air space or
- Some other method of preventing precipitation from reaching the inner leaf. An external cavity wall may consist of the following:
- An outside leaf of masonry (brick, block, natural or reconstructed stone).
- Minimum 50 mm uninterrupted cavity. Where a cavity is bridged (by a lintel etc.), a damp-proof course or tray should be inserted in the wall so that the passage of moisture from the outer to the inner leaf is prevented. This is not necessary where the cavity is bridged by a wall tie or where the bridging occurs, presumably, at the top of a wall and is then protected by the roof. Cavity walls may also be bridged by cavity barriers and fire stops where other parts of the Building Regulations require this (such as Part B or Part E). Where an opening is formed in a cavity wall, the jambs should have a suitable vertical damp-proof course or the cavity should be provided with a suitable cavity closure so as to prevent the passage of moisture.
- An inside leaf of masonry or framing with suitable lining.
- To ensure structural robustness and weather resistance in the wall, masonry units should be laid on a full bed of mortar and the cross joints should be continuously and substantially filled.
- Where a cavity is only partially filled with insulation, the remaining cavity should be at least 50 mm wide (Figure 8.18).

The weather resistance and thermal insulation performance requirements for all wall types are summarised in Figure 8.18.

Durability

According to Regulation 7 of the Building Regulations, building work must be carried out:

- With proper and adequate materials which are
  - appropriate for the circumstances in which they are used and

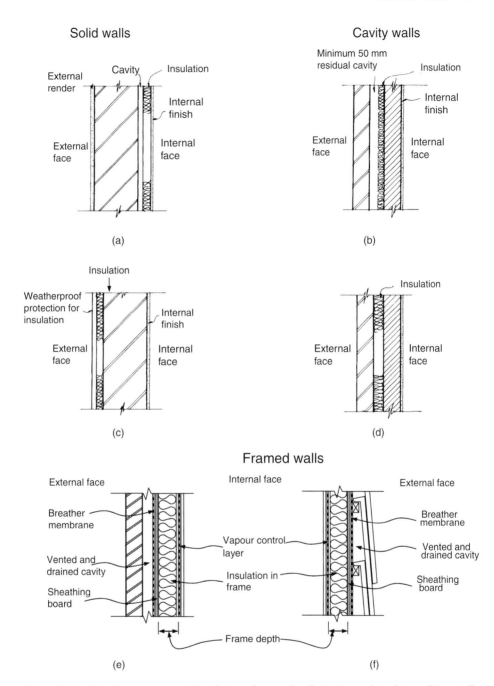

Figure 8.18 Weather resistance and insulation of external walls: (a) internal insulation, (b) partially filled cavity, (c) external insulation, (d) fully filled cavity, (e) brick clad timber framed wall and (f) tile clad timber-framed wall.

o adequately mixed and prepared, and applied, used or fixed so as adequately to perform the functions for which they are designed.
■ In a workmanlike manner.

Therefore, materials should be suitable in nature and quality in relation to the purpose for which, and the conditions in which, they are used.

The regulations do not specify precise durability standards in relation to materials. However, they do discuss in general terms the conditions under which a material might be considered to be unsuitable due to its short-lived nature when compared to the expected life of the building. Therefore, if a material or a component is inaccessible for replacement should it fail, and its failure would create a serious health risk, then it is unlikely that the material or component would be suitable or would satisfy the building regulation standards of fitness for purpose.

Fire safety

For external walls, the most important fire safety considerations relate to fire resistance, the ability of the external surfaces to transmit fire across their surfaces and the number and position of structural openings for doors and windows relative to the site boundaries.

Fire resistance is measured in minutes and is related to the period that the wall would fail if a fire developed inside (or sometimes outside) the building. For extensions to dwellings covered by this book, the maximum period of fire resistance that has to be achieved is 30 min. This has to be achieved from both sides of the wall if it is within 1 m of the boundary, but only from the inside if the wall is more than 1 m from the boundary. Fire resistance is important for a number of reasons:

■ It prevents the building from collapsing quickly in the event of a fire and gives the occupants a chance to escape.
■ It gives fire fighters a period of time to carry out a rescue.
■ It prevents the spread of fire to another building close by.
■ It prevents the building from being engulfed by fire should a neighbouring building be ablaze.

All of the wall constructions shown in this book will, under most ordinary circumstances, provide adequate fire resistance, although there are certain restrictions where your extension is close to your neighbour's boundary. The controlling critical dimension is 1 m. Below this distance you will be restricted in the amount of combustible claddings (such as softwood boarding) and wall openings that you can have in such a wall.

The simplest way of checking that you comply with the fire rules for external walls is to use Method 1 shown in Approved Document B Volume 1, which is illustrated in Diagram 22 in the Approved Document (Table 8.4).

This method applies only to dwelling houses, flats, maisonettes or other residential buildings, which

■ Are not less than 1 m from the relevant boundary,
■ Are not more than three storeys high (basements not counted) and
■ Have no side which exceeds 24 m in length.

Table 8.4 Permitted unprotected areas for Method 1: Approved Document B Volume 1, Diagram 22.

Minimum distance (A) between side of building and relevant boundary (m)	Maximum total area of unprotected areas (m^2)
1	5.6
2	12
3	18
4	24
5	30
6	No limit

Therefore, the permitted limit of unprotected area (i.e. windows, doors and combustible claddings) in an external wall varies according to the size of the building and the distance of the side from the relevant boundary. Any parts of the side in excess of the maximum unprotected area should have the recommended fire resistance. Certain small areas may be discounted. These are as follows:

- Any unprotected area of not more than 0.1 m^2 which is at least 1.5 m away from any other unprotected area on the same side of the building.
- One or more unprotected areas, with a total area of not more than 1 m^2, which is at least 4 m away from any other unprotected area on the same side of the building or compartment, except a small area of not more than 0.1 m^2 as described above.
- Where part of an external wall is regarded as an unprotected area merely because of combustible cladding more than 1 mm thick, the unprotected area presented by that cladding is to be calculated as only half the actual cladding area.

So any of the small areas referred to here can be closer than 1 m to the boundary (and can, in fact, be actually on the boundary), although a window situated on a boundary would have to be non-opening and obscure glazed under the planning laws mentioned in Chapter 2.

An another requirement for cavity external walls is worthy of mention. Since cavities in walls can be conduits for the passage of smoke and hot gases, they need to be stopped with non-combustible construction:

- Around openings for doors and windows where the frame itself could be considered to be an adequate fire stop unless it was made from PVC.
- At the junction between the external wall and the party wall between neighbouring buildings and at their tops unless the wall is of non-combustible masonry construction. In this case, the wall would have to be totally filled with insulation to avoid having a fire stop at its head.

FURTHER INFORMATION

References used in this chapter include the following:

- Building Regulations, Approved Document A – Structure
- Building Regulations, Approved Document B – Fire Safety, Volume 1, Dwellinghouses

- Building Regulations, Approved Document C – Site Preparation and Resistance to Contaminants and Moisture
- Building Regulations, Approved Document L – Conservation of Fuel and Power, Volume L1B – Conservation of Fuel and Power in Existing Dwellings
- Building Regulation 7, Materials and Workmanship

Further information can be obtained from

http://www.uktfa.com/
http://www.sips.uk.com/
http://www.icfinfo.org.uk/

9 Internal walls

Questions addressed in this chapter:

How can I construct an internal wall?
Are there any new materials and methods in use for these constructions?
What Building Regulations apply to these construction areas?

PARTITION WALLS

Internal walls can be categorised into three different types, namely

- Non-load-bearing partition walls;
- Load-bearing partition walls and
- Compartment walls (sometimes called separating walls or party walls).

Partition walls can be described as walls which divide up the internal space of a building. Non-load-bearing partition walls will simply support their own weight, as opposed to load-bearing partition walls, which will assist in distributing the loads of the building (e.g. floors, roof etc.) safely to the foundations. Therefore, load-bearing partition walls must have their own independent foundation. Non-load-bearing partition walls can be built straight off the oversite concrete slab or the first floor construction in two storey extensions.

Compartment walls are internal walls that separate the internal area in the same way as partition walls. However, these walls serve as a means of protection from fire. Therefore, compartment walls can be used to provide a fire-protected stairwell as a means of escape or as a wall between two different dwellings, such as the wall between a pair of semi-detached houses.

Partition walls are traditionally constructed using two methods. Concrete blocks are used for ground floor partitions and timber stud partitions are used on all floors above this. It is also possible to have stud partitions where light steel framing is used instead of timber. Additionally, stud walls can be used on the ground floor, depending on personal choice and other methods of construction adopted. Also, if a partition wall is load bearing the concrete block construction can extend on to the first floor accommodation, as this wall may provide support to the roof structure.

Extending and Improving Your Home: An Introduction, First Edition. M.J. Billington and C. Gibbs.
© 2012 M. J. Billington and C. Gibbs. Published 2012 by Blackwell Publishing Ltd.

Blockwork partition walls

Concrete blocks are used for partition walls for several reasons:

■ They are more economical than lightweight insulation blocks.
■ They offer better sound insulation owing to their mass.
■ Thermal insulation is not required, as partition walls separate habitable rooms.

The concrete block can be finished in numerous ways (further information on finishes is given in Chapter 11).

Partition walls must be tied into external walls. This can be achieved in several ways.

Indents

Indents or toothings can be left in the external wall at the location of the partition wall and the blockwork partition can then be tied in by building the blockwork into the external wall into these indents (Figure 9.1). Whilst this provides an excellent tie it is now not the most recommended because the partition wall and the external wall may be of different materials, so will expand and contract (through temperature changes and moisture content) at different rates, which can cause cracking (also referred to as differential movement). The concrete blocks can also cause cold bridging through the thermal performance of the external wall. As the partition wall is constructed off the oversite concrete slab and the external wall is constructed off a separate concrete foundation it is also possible that differential structural movement and, therefore, cracking along the junction of the two walls may occur.

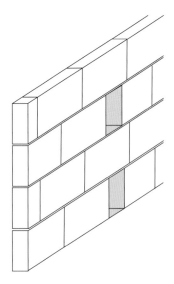

Figure 9.1 Indents or toothings in the external wall at the location of the partition wall.

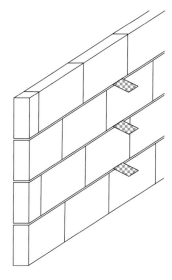

Figure 9.2   Metal ties in the mortar joints.

Metal ties

During construction of the external wall, metal ties (wall ties or expanded metal lathing) can be built into the mortar joints to provide the tie for the partition wall construction (Figure 9.2). The concrete blocks are, therefore, not built into the external wall, so some of the problems associated with indents are eliminated, as the ties will allow for some movement without having a detrimental effect on either the partition or the external wall.

Helical ties

If the external wall (or internal leaf of cavity wall) is constructed of a lightweight block then these spiral bars can be hammered into position during the construction of the partition (Figure 9.3). No separate provision needs to be made in the construction of the external wall, so erection is quicker and there is less risk of forgetting to insert the ties or from incorrect setting out. From a health and safety point of view, there are no protrusions or pieces of metal sticking out from the wall for the workforce to be injured.

Timber stud partition walls

Partition walls can also be constructed from timber; these are called timber stud partition walls (Figure 9.4). The timber framework is constructed from softwood and can either be

Figure 9.3   Helical ties can be hammered into position.

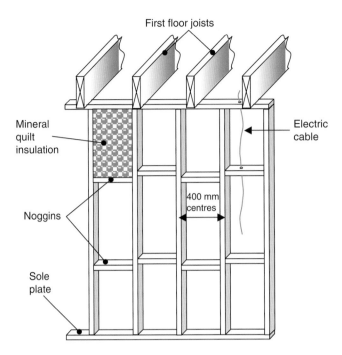

Figure 9.4　Timber stud partition.

pre-made and fixed into position or can be constructed *in situ* (in the place where it is to remain). The vertical uprights are positioned at 400 mm centres to ensure that standard boards can be fitted to the framework without unnecessary cutting and waste. The void within the timber framework can be filled with mineral wool insulation if an increase in sound insulation or thermal insulation is required. In dwellings this is a Building Regulation requirement for the following partition walls:

- A wall between a bedroom and any other room in the house.
- A wall between a bedroom and a room containing a water closet (but not if the room containing the water closet is en suite with the bedroom).

The requirement can be met for a stud partition by fitting a minimum 25 mm rockwool or fibreglass quilt with a density of at least $10 \, \text{kg/m}^3$ in the cavity.

The void also provides a sensible location to duct services (e.g. electrical cables and heating and hot and cold water pipes).

The stud frame is fixed to the oversite concrete and the first floor joists if located on the ground floor or fixed to the first floor joists and the ceiling joists if on first floor. The partition is also secured to the external wall by a variety of different fixings (e.g. bolts, screws and angle plates).

Partitions can also be constructed from metal profiles. Metal channels are fixed to both floor and ceiling then vertical members are cut and fixed between these two channels (Figure 9.5). Either side of this partition frame can then be covered with plasterboard that is fixed to the framework using self-tapping dry wall screws.

These types of partition are common in commercial/industrial situations and are now becoming more common in residential situations because they are quick and economical to install.

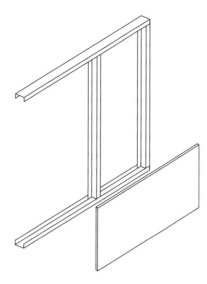

Figure 9.5   Metal frame partition.

BUILDING REGULATIONS AND APPROVED DOCUMENT GUIDANCE

There are very few Building Regulation requirements that apply to the design and construction of internal partitions. The principal ones include

- Sound insulation (covered above and in Approved Document E – Resistance to the Passage of Sound);
- Structural stability (Approved Document A – Structure) and
- Safety in fire (Approved Document B – Fire safety, Volume 1 – Dwellinghouses).

Structural stability

If the partition walls are of masonry construction (concrete blocks or bricks) and are load bearing, they must have a thickness of at least 100 mm (excluding the thickness of any applied finish).

Additionally, it is often the case that load-bearing internal partition walls provide lateral stability to the external walls of a building. Such walls are referred to as buttressing walls and must comply with the following rules:

- One end bonded or securely tied to the supported wall.
- The other end bonded or securely tied to another buttressing wall, pier or chimney.
- No opening or recess greater than 0.1 m^2 in area within a horizontal distance of 550 mm from the junction with the supported wall, and openings and recesses generally disposed so as not to impair the supporting effect of the buttressing wall.
- A length of not less than one sixth of the height of the supported wall.
- A minimum thickness of 100 mm (Figure 9.NaN).

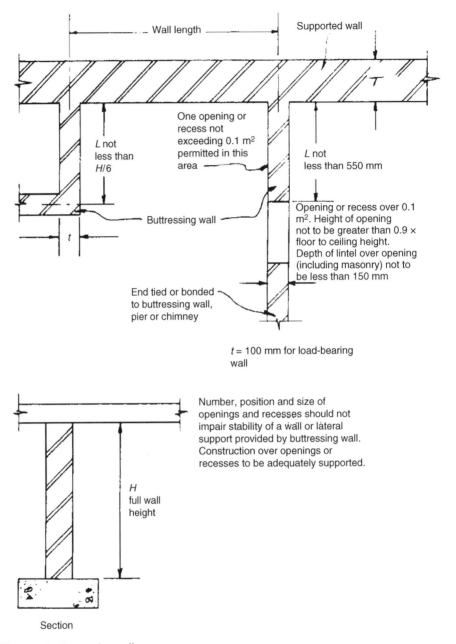

Figure 9.6 Buttressing walls.

Stud walls are not usually considered to be load bearing. This is possible but if this was the case they would have to be designed by a structural engineer in accordance with the recommendations of BS 5268* Structural Use of Timber. For light steel partitions the relevant design code is BS 5950* Structural Use of Steel in Buildings.

*Reference to BS5268 and BS5950. These standards were superseded by the relevant Eurocode standards in 2010 but are still referred to in Approved Document A so may be used alongside the national standards until these are withdrawn.

Safety in fire

Internal partitions in dwellings only need to be fire rated if they enclose a protected stairway. Normally, there is only a need for a protected stairway in a three storey dwelling and it is possible for two storey buildings to have an open plan ground floor, meaning that there would be no walls surrounding the stairway. It should be born in mind that if alterations are carried out to a three storey dwelling then the partitions surrounding the stairway should not be removed or have their fire resistance reduced below 30 min. Thirty minutes fire resistance can be achieved for a timber stud partition by constructing the partition with studs at least 40 mm thick, spaced at no greater than 600 mm centres and clad each side with 12.5 mm plasterboard with the joints taped and filled.

FURTHER INFORMATION

References used in this chapter include the following:

- Building Regulations, Approved Document A – Structure
- Building Regulations, Approved Document B – Fire Safety, Volume 1, Dwellinghouses
- Building Regulations, Approved Document E – Resistance to the Passage of Sound

Further information can be obtained from

http://www.lafargeplasterboard.co.uk
http://www.interior-facility.com/

10 Roofs

Questions addressed in this chapter:

How can I construct a roof?
What materials can I use for the roof covering?
What design issues do I need to consider?
Are there any new materials and methods in use for these constructions?
What Building Regulations apply to these construction areas?

ROOF STRUCTURE AND COVERINGS

The roof of an extension to a house is one of the main features that clearly illustrates good design. The roof can reflect the character of the house to which the extension is attached like no other feature. It can be dominant with a steep pitch and large overhanging eaves or can be an insignificant flat roof hidden behind a parapet wall. Apart from pitch, the other major feature of a roof is the covering material; for this there is a huge range of possible coverings of all imaginable colours and textures. It is possibly true that most people prefer a pitched roof but it may not always be the best solution to the many issues that face the designer.

Whichever form of roof is desired there are a certain number of common factors that all roofs must satisfy. These are

- The ability to carry loads and distribute them safely to the walls and thus to the foundation;
- The provision of an acceptable level of thermal insulation;
- The provision of an acceptable level of sound insulation;
- The ability to keep out the elements of the weather;
- A standard of durability suitable for the life of the building;
- Structural stability;
- The provision of an appropriate level of fire resistance and
- An appropriate aesthetic appearance (especially in conservation areas and when dealing with listed buildings).

Most of these requirements are covered by Building Regulations (detailed at the end of this chapter) and other legislation (Chapter 2). However, it is interesting to note that whilst a degree of sound insulation is desirable in roofs, so as to provide a satisfactory and reasonably quiet internal environment, current building regulations do not cover sound penetrating from outside sources, although they do address the issue of sound being transmitted from one dwelling to another via the intervening roof space, such as is found in semi-detached and terraced properties.

Extending and Improving Your Home: An Introduction, First Edition. M.J. Billington and C. Gibbs.
© 2012 M. J. Billington and C. Gibbs. Published 2012 by Blackwell Publishing Ltd.

Pitched roofs

When the external wall has been constructed to the specified height, in preparation for the roof construction it is usual for it to be provided with a timber wall plate. It is, therefore, necessary to take the thickness of this wall plate into consideration when building the wall. The wall plate is normally constructed from $100 \times 50 \, mm^2$ softwood timber (although, wall plates larger than this can also be used) and is laid on a mortar bed to ensure it can be laid level, eliminating any inaccuracies in the wall. If joins are to be made in the wall plate, a lap joint is recommended.

The wall plate provides a means of fixing the roof construction to the external wall structure. It is essential therefore that this is level, parallel and square with the opposite side of the structure where another wall plate would be laid. To prevent the risk of roof damage through wind uplift, the wall plate is secured to the external wall by the use of mild steel anchor straps. This strapping of the roof to the walls also provides valuable lateral stability to the tops of the external walls. These straps are commonly fitted after the roof structure so as not to coincide with any rafters, ceiling joists or trusses and must be no more than 2 m apart along the wall length (Figure 10.1).

There are two common forms of construction for a pitched roof – a cut roof and manufactured trussed rafters. With a cut roof all work is done on site. This includes setting out, cutting of timbers and construction of the roof structure (Figure 10.2). Manufactured trussed rafters, on the other hand, are designed and manufactured in a factory. These are then delivered to site and simply erected in position, fixed down and provided with wind bracing.

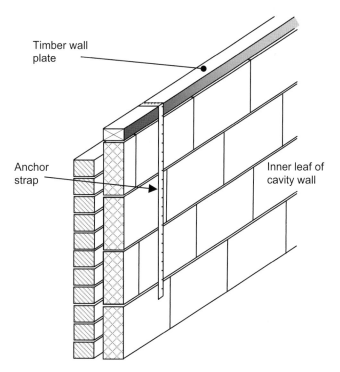

Figure 10.1 Positioning and fixing of a wall plate.

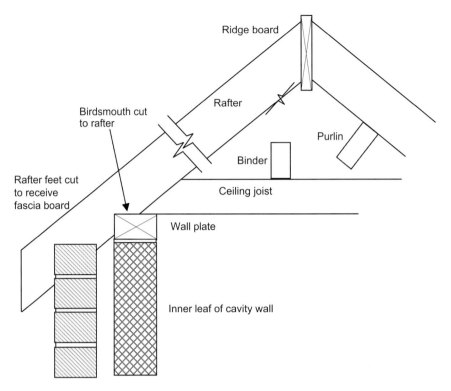

Figure 10.2   Arrangement of roof timbers.

A cut roof is very labour intensive on site and requires larger sections of timber for the roof construction. The section sizes of roof members (rafters, ceiling joists, purlins and binders) are determined by the

- Proposed loadings;
- Spacing between the rafters;
- Span between supports;
- Strength class of the timber and
- Economics, since some structural members may be able to span the required distance but may be of an unusual and therefore expensive size.

By applying this information to tables given in '*Span Tables for Solid Timber Members in Floors, Ceilings and Floors (excluding trussed rafter roofs) for Dwellings*' produced by TRADA Technology Ltd, a section size can be determined, although most competent carpenters have rules of thumb by which they can judge the sizes of the various roof members, provided that the loading and spans are not out of the ordinary.

A cut roof consists of sloping members called rafters that will be cut to sit on, and be skew nailed to, the wall plate using a birdsmouth cut (i.e. a V-shaped cut in the underside of the rafter that allows it to fit snugly on the wall plate). However, a proprietary truss clip is a preferred option as this reduces the risk of damage to the rafter (Figure 10.3). The rafter is also cut at an angle at the top where it joins the ridge board.

In addition, ceiling joists will also be required to extend from the rafter feet at wall plate level to the rafter at the other side of the building. The ceiling joist satisfies two requirements. Firstly, it provides a framework to fix a ceiling finish to and, secondly,

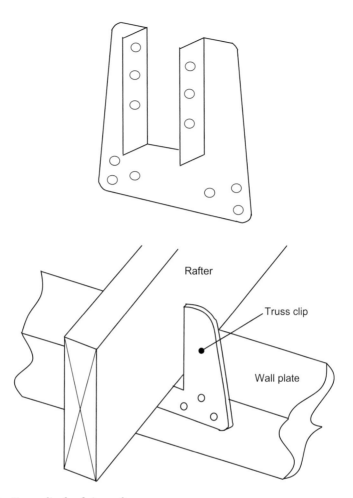

Figure 10.3 Truss clip for fixing rafters.

it prevents the spread of the rafter when loads are applied (e.g. dead loads and imposed loads), since it is securely fixed to the rafter feet on each side of the building. If the span of the roof from one side of the building to the other is large (i.e. more than about 4 m) the ceiling joist will need intermediate support from either an internal wall or a timber beam called a binder which runs at right angles to the ceiling joists. A similar type of support may also be needed for the rafters if they have to span a large distance. This is normally provided by a timber beam called a purlin. These are illustrated in Figure 10.2.

A manufactured trussed rafter incorporates all the necessary timbers in a single unit (Figure 10.4). These are becoming very cost effective and can considerably increase the speed of the roof structure construction. When being lifted into position, each trussed rafter should be stored and lifted by its loading points (i.e. where the ceiling tie meets the rafter section) and should always be kept vertical. Inappropriate storage, lifting and twisting of manufactured trussed rafters can result in the plate fixings becoming loose and, therefore, the required design level cannot be achieved and failure may result. When lifted into position onto the wall plate, the trussed rafters are fixed into position using truss clips (Figure 10.3) and no cutting is required. However, manufactured trussed rafters require sufficient bracing to ensure their structural integrity. These include horizontal bracing at

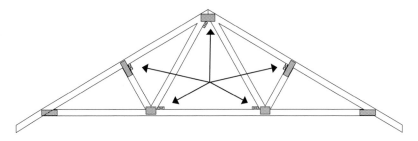

Figure 10.4   Section and plan of trussed rafter bracing.

ceiling tie level and other node points and also diagonal bracing from ridge to wall plate (Figure 10.5) to resist lateral wind forces.

If a gable wall has been provided (a gable is that triangular piece of masonry that extends the end wall of the house up to the ridge), a gable ladder will need to be constructed to extend the roof structure over the gable wall to ensure adequate weather exclusion by extending the roof covering over the top of the gable brickwork.

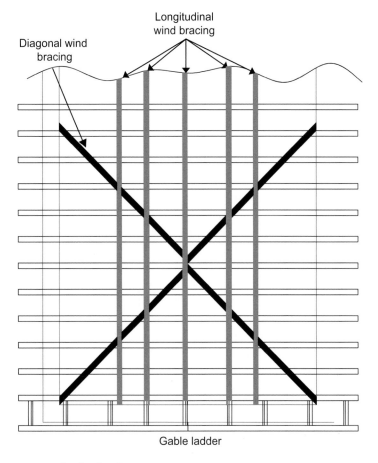

Figure 10.5   Trussed rafter bracing.

Once the roof structure has been completed this can then receive the appropriate roof covering and the development soon takes on a new lease of life when you have a dry working environment and the structure can start to dry out.

Flat roofs

Flat roofs can be seen in numerous locations, these are not normally the primary choice of roof. Whilst the flat roof structure can be designed adequately to last a considerable time, flat roof coverings are renowned for having a reduced life span in comparison to traditional pitched roof coverings. The structure of a flat roof will consist of horizontal joists spanning from wall plate to wall plate.

The section size of these joists is determined by the

- Proposed loadings;
- Spacing between the joists;
- Span between supports and
- Strength class of the timber.

By applying this information to tables given in '*Span Tables for Solid Timber Members in Floors, Ceilings and Floors (excluding trussed rafter roofs) for Dwellings*' produced by TRADA Technology Ltd, a section size can be determined.

Fixed to the top of these joists will be tapered 'firring' pieces. These 'firring' pieces provide a slight fall in the roof covering to ensure adequate disposal of rainwater. The recommended fall should be not less than 1 in 80 and most sources recommend a fall of not less than 1 in 40 to ensure that roof water drains from the roof surface to the rainwater outlets.

The roof decking of roofing grade plywood (i.e. water and boil proof [WBP]) at least 18 mm thick or softwood tongued-and-grooved boarding at least 25 mm thick sits on top of these firring pieces providing a solid surface to take the chosen roof covering.

Traditionally, insulation is laid between the joists to achieve a sufficient U-value. There should also be a space between the insulation and the underside of the roof decking in order to provide adequate ventilation (Figure 10.6). Ventilation in a flat roof is essential to ensure that condensation does not occur, as this inevitably causes rotting of timbers. This is more prevalent in flat roofs as opposed to pitched roofs because there is no visual

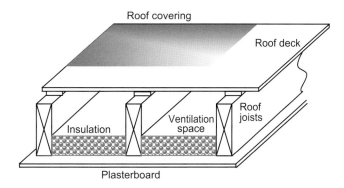

Figure 10.6   Flat roof construction – cold deck.

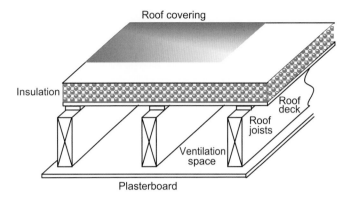

Figure 10.7   Flat roof construction – warm deck.

access to the structure of a flat roof to monitor any such problems and the ventilated space is smaller than in a pitched roof, so less able to hold water in a vapour form. Therefore, significant damage can be caused if a defect such as this occurs in a flat roof.

When the insulation is placed between the joists and sits on the ceiling finish (e.g. plasterboard), this is referred to as a 'cold deck roof' (Figure 10.6). Condensation problems are more likely in this type of construction and the use of foil-backed plasterboards or a vapour control layer placed between the plasterboard and insulation is essential.

An alternative to this is to locate the insulation on top of the roof decking. This is referred to as a 'warm deck roof' (Figure 10.7). The roof covering will then be placed on top of the insulation, so the choice of insulation product must be carefully considered. There is less risk of condensation occurring in the roof structure if the insulation is moved further toward the external part of the structure. However, ventilation can still be provided as an extra precaution. Again, it is essential that a vapour control layer is placed on the warm side of the insulation.

ROOF COVERINGS

Pitched roof coverings

Pitched roof coverings commonly consist of an underfelt, fixing battens and the covering element itself.

Reinforced bitumen-based underfelts have been used for a considerable number of years. However, modern technology now sees the common use of plastic-based underfelts. Plastic underfelts can be easier to use than bituminous felts and also easier to repair should any damage occur. They can also be microporous so that they are breathable. This allows any residual condensation that forms in the roof space to be ventilated out through the felt layer but does not allow any liquid water that might penetrate the tiles or slates to pass through the felt into the roof space. The felt is laid horizontally across the roof structure and is nailed to the rafters. A minimum of 150 mm lap should exist between adjacent layers of underfelt. Both sides of the pitched roof are covered in this way with a

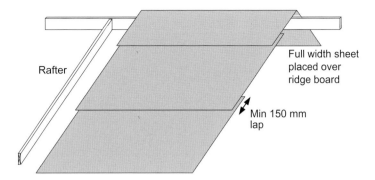

Figure 10.8 Roofing underfelt.

single complete roll of felt being laid over the ridge (Figure 10.8). The first row of underfelt should also protrude enough past the fascia so that it will turn into the rainwater gutters when they are fixed. This protrusion helps protect the fascia, soffit and rafters as well as reducing the amount of rain falling on the wall below (Figure 10.9). There are proprietary plastic trays that could be used instead of allowing for the protruding underfelt but of course this must be specified and fitted before the final roof covering is installed. It is usual for a small triangular piece of wood (called a tilting fillet) to be placed on the end of each rafter to support the underfelt and prevent it sagging where it passes over the fascia board. Sometimes this is achieved by using a continuous strip of thin plywood or fibre cement sheet.

It is important to have the roof covering specified and designed prior to fixing the underfelt as battens will need to be fixed immediately after the underfelt and this will reduce the risk of any damage by the wind during construction. The position of the battens

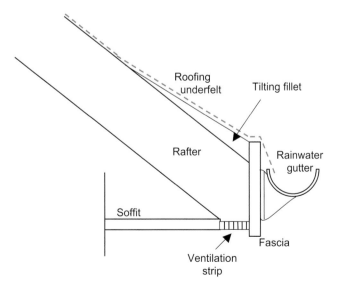

Figure 10.9 Eaves detail showing underfelt protruding into gutter.

is determined by the chosen roof covering. There is a large array of roof coverings available for pitched roofs, the more common being

- Clay tiles;
- Concrete tiles;
- Man-made slate and
- Natural slate.

The choice of roof covering could be decided from personal choice. However, this will have to be agreed with the planning authority. This will be quite specific should the property be in a conservation area or, if it is a listed building, the planning authority may then stipulate the exact type of roof covering to be used.

Owing to the large number of sizes and types of roof coverings available, the roof covering must be designed in order that the location of the treated (sometimes referred to as tantalised or protimised in the trade) fixing battens can be determined. Different roof coverings have different specifications and the required minimum pitch for the covering must be determined prior to the roof construction. Not all roof tiles and slates can be laid at the same pitch and for some low-pitched roofs (less than 12.5°) there may be a very small range of suitable roof coverings.

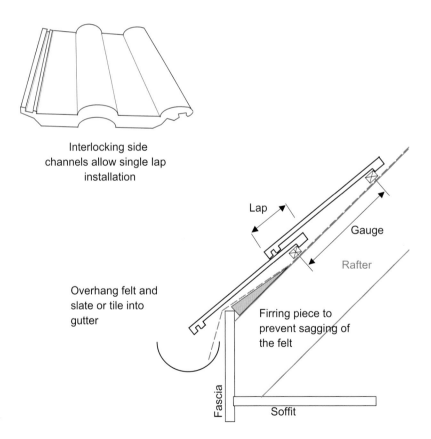

Figure 10.10   Single lap interlocking tile roof covering.

There are several items to be taken into consideration when designing the roof covering:

- Ensure adequate overhang of the roof coverings into the rainwater gutter;
- The length of the roof covering unit (e.g. tile, slate etc.);
- The headlap and
- The length of the rafter.

Interlocking tiles, whether of concrete or clay, allow for single lap tiling, whereby the tile above only laps over one tile underneath. This is possible due to the interlocking side channels protecting the junctions between each tile (Figure 10.10).

However, slates and clay tiles that are not interlocking will allow water to pass through the gaps between each slate/tile. Therefore, the slate above laps over two slates below to ensure these gaps are covered and rainwater is kept out of the building (Figure 10.11).

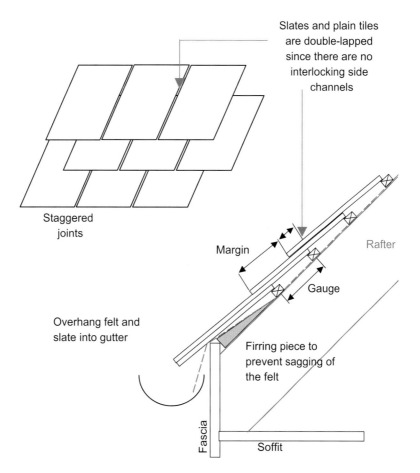

Figure 10.11 Slate roof covering.

Flat roof coverings

Traditionally, flat roofs were covered in three layers of built-up bituminous felt. The felt consisted of a bitumen-impregnated mat which was made from vegetable fibres (often hemp). Over the years these types of felt have been supplanted by a new breed of felts which give greater lifespan, elasticity and wear resistance than the traditional types.

The main changes have been in the reinforcing mat, which is now made from glass or polyester fibres, and in the impregnation materials, which are now modified with chemicals such as styrene butadiene styrene (SBS) to give a fatigue resistance 20 times better than that obtained using traditional felts.

Such felts are built up in three layers with overlapping staggered joints, each layer being bonded to the one underneath with liquid bitumen. The disadvantages of such systems are that they are comparatively inflexible, especially if they have to be formed to fit round tight angles, and the bitumen that sticks the sheets together has to be heated to melt it before it can be applied. Because of this a large number of other flat roofing materials have been developed that are cold applied and, therefore, safer to use. In addition, they can be applied in a single layer (called single ply membranes) and are easier to apply, especially where complicated details must be formed. These modern single ply membranes are based on man-made materials derived from the oil industry and are not considered to be sufficiently 'green' by some people. They do, however, have significant advantages over traditional flat roof felts in that they are longer lived and more flexible but are not always as robust and may be more expensive to install. There are also some single ply materials based on recycled polymer modified rubber.

The other main areas of flat roof coverings are sheet metal and liquid applied coatings.

Lead is the traditional sheet metal flat roof covering that has been used for hundreds of years. Today it is regarded as hazardous, expensive and rather heavy, so it tends only to be used to repair or replace flat roof coverings on listed buildings.

Liquid applied coatings include traditional asphalt roofing, which has to be applied hot so is relatively hazardous, and the new range of plastic and resin based spray-on coatings, known affectionately in certain parts of the roofing industry as 'spray and pray'. This is rather unfair as such coatings have the advantages of being cold applied and able to accommodate extremely awkward roof shapes and details. Some can also be applied by squeegee or roller when it is usual to prime the roof deck and then saturate a polyester or glass fibre mat with the liquid material to form a membrane. Such systems can also be particularly durable.

OTHER COVERINGS

Many other coverings exist that can be used depending on your particular situation but are not being covered by this text, these include

- Thatch
- Wood shingles
- Fibreglass
- Glass

- Steel
- Copper
- Lightweight metal roof tile sheets
- Corrugated fibre cement and metal roof sheets

BUILDING REGULATIONS AND APPROVED DOCUMENT GUIDANCE

At the start of this chapter the functional and performance requirements of roofs were referred to. From the items listed, building regulations apply to the following:

- The ability to carry loads and distribute them safely to the foundation and provide an adequate level of structural stability (Part A – Structure).
- The provision of an acceptable level of thermal insulation (Part L – Conservation of Fuel and Power).
- The ability to keep out the elements of the weather and avoid excessive condensation – (Part C – Site Preparation and Resistance to Contaminants and Moisture).
- A standard of durability suitable for the life of the building (Regulation 7).
- The provision of an appropriate level of fire resistance (Part B – Fire Safety).

Structural stability

Part A has already been discussed in Chapters 5 and 8, where the requirements for foundations and walls, respectively, were looked at. Approved Document A considers roof structures for small extensions from the following viewpoints:

- What loads must be taken by the roof structure?
- What ways can we design a roof structure to carry these loads and transmit them to the walls so that the walls and roof act together as a structural system?
- What materials are we using to carry these loads?

In the sections above we have considered both pitched and flat roof structures constructed in timber and have suggested designs and methods of construction that will meet the requirements of the regulations. It is, of course, possible to use both steel and concrete to construct roof structures but these materials are relatively uncommon in domestic extension roofs and are beyond the scope of this book. If it is desired to construct the roof structure in steel or concrete, the services of a qualified structural engineer will be needed.

DESIGN FOR THERMAL EFFICIENCY

Part L of the Building Regulations covers the construction of extensions to dwellings and gives guidance on the thermal performance expected in Approved Document L1B, Conservation of Fuel and Power in Existing Dwellings. For simple designs without excessive window and door areas the guidance is quite straightforward. However, where the area of windows and doors exceeds 25% of the floor area of the extension it is necessary

Table 10.1 Extract from Table 2 of Approved Document L1B – standards for new thermal elements.

Element	U-value standard (W/m²K)
Wall	0.28
Pitched roof – insulation at ceiling level	0.16
Pitched roof – insulation at rafter level	0.18
Flat roof, or roof with integral insulation	0.18
Floors	0.22
Window, roof window or rooflight	1.6*
Doors	1.8

*or Window Energy Rating (WER) of Band C or better.

to carry out a thermal calculation and it is wise to use the services of an expert to do this, as it involves complicated calculations and an understanding of the thermal properties of the materials used.

Firstly, the thermal elements of the extension (roof, walls, ground floor, external windows and doors) must achieve certain thermal values termed U-values (a U-value is called a thermal transmittance coefficient and the larger the number the worse the standard of performance).

The U-values that must not be exceeded are shown in Table 2 from Approved Document L1B (an extract from this table is shown in Table 10.1).

Secondly, to avoid having to carry out thermal calculations on the extension, the area of new windows and doors should be restricted to no more than 25% of the newly created extension floor area. This area can be increased if, as a result of the extension, windows and doors that were in the original house are covered by the extension. So the areas of such windows and doors can be added to the 25% floor area allowance without incurring the need for a thermal calculation.

Weather resistance and avoidance of condensation

The roof of a building should

- Resist the penetration of precipitation to the inside of the building;
- Not be damaged by precipitation;
- Not transmit precipitation to another part of the building that might be damaged;
- Be designed and constructed so as not to allow interstitial condensation to adversely affect its structural and thermal performance and
- Not promote surface condensation and mould growth under reasonable occupancy conditions.

Approved Document C contains few practical recommendations for the design and construction of roof coverings, preferring instead to merely make reference to a series of codes of practice and standards which are routinely used by good roofing companies. This trend of continually referring to additional sources of guidance is making the Approved Documents much less useful to the practitioner as, in many cases, they can only be regarded as a directory and not as a direct source of technical guidance in their own right.

To provide more useful information on controlling condensation in roofs (which can be a serious problem) the following information has been put together from a number of different sources.

When condensation occurs in roof spaces it can have two main effects:

- The thermal performance of the insulant materials may be reduced by the presence of the water.
- The structural performance of the roof may be affected due to increased risk of fungal attack.

Approved Document C recommends that interstitial condensation in roofs should be limited such that the thermal and structural performance of the roof will not be adversely affected.

It should be noted that the provisions of Approved Document C apply to roofs of any pitch even though a roof which exceeds 70° in pitch is required to be insulated as if it were a wall. Additionally, small roofs over porches or bay windows and so on may sometimes be excluded from the requirements if there is no risk to health or safety.

In swimming pools and other buildings where high levels of moisture are generated, there is a particular risk of interstitial condensation in walls and roofs because of the high internal temperatures and humidities that exist. In these cases specialist advice should be sought when these are being designed.

For cold deck roofs (where the insulation is placed at ceiling level and can be permeated by moisture from the building) the requirements of the regulations can be met by the provision of adequate ventilation in the roof space. In such roofs it is obviously essential that moist air is prevented from reaching the roof space where it might condense within or on the insulation layer; this is particularly important above areas of high humidity such as bathrooms and kitchens. Weak points in the construction where this might occur include

- Gaps and penetrations for pipes and electrical wiring (these should be filled and sealed) and
- Loft access hatches (where an effective draught seal should be provided to reduce the inflow of warm moist air).

Further, very sound guidance may also be found in the Building Research Establishment Report BR 262 *Thermal insulation: avoiding the risks*.

The following guidance covers a few typical examples of commonly found forms of construction where the insulation is placed at ceiling level (cold deck roofs).

Roofs with a pitch of 15° or more

Pitched roofs should be cross-ventilated by permanent vents at eaves level on the two opposite sides of the roof, the vent areas being equivalent in area to a continuous gap along each side of 10 mm width. Ridge ventilation equivalent to a continuous gap of 5 mm should also be provided where the pitch exceeds 35° and/or the spans are in excess of 10 m.

Mono-pitch or lean-to roofs should have ventilation at eaves level as above and also at high level, either at the point of junction or through the roof covering at the highest practicable point. The high level ventilation should be equivalent in area to a continuous gap 5 mm wide (Figure 10.12).

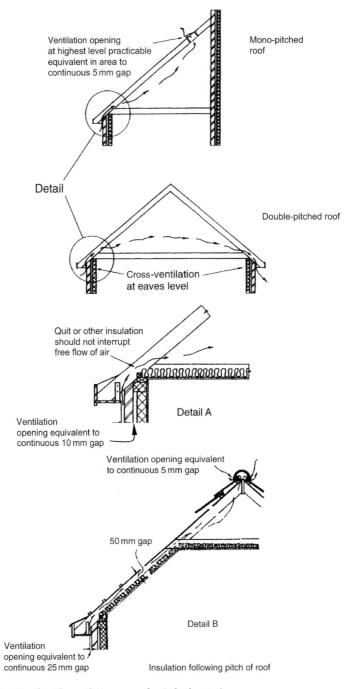

Ventilation opening
at highest level practicable
equivalent in area to
continuous 5 mm gap

Mono-pitched
roof

Detail

Double-pitched roof

Cross-ventilation
at eaves level

Quit or other insulation
should not interrupt
free flow of air

Detail A

Ventilation
opening equivalent to
continuous 10 mm gap

Ventilation opening equivalent
to continuous 5 mm gap

50 mm gap

Detail B

Ventilation
opening equivalent to
continuous 25 mm gap

Insulation following pitch of roof

Figure 10.12   Roof void ventilation – roofs pitched at 15° or more.

In recent years, vapour permeable underlays have come onto the market. If these are used without continuous boarding below the tiling battens it is not necessary to ventilate the roof space below the underlay. However, where boarding is provided a ventilated space should be formed above the underlay by using 25 mm battens and counter-battens, and provision of ventilation at low level equivalent to a 25 mm continuous gap and high level equivalent to a 5 mm continuous gap. Simply relying on fortuitous ventilation through the tile/slate joints should not be relied on to ventilate this space adequately.

Roofs with a pitch of less than 15°

In low-pitched roofs the volume of air contained in the void is less and, therefore, the risk of saturation is greater.

This also applies to roofs with pitch greater than 15° where the ceiling follows the pitch of the roof. The high level ventilation should be equivalent in area to a continuous gap 5 mm wide.

Cross-ventilation should again be provided at eaves level but the ventilation gap should be increased to 25 mm width (Figure 10.13).

Resistance to surface condensation and mould growth in roofs

To resist surface condensation and mould growth in roofs and roof spaces, care should be taken to design the junctions between the elements and the details of openings, such as windows, so that thermal bridging and air leakage is avoided. This can be done by following the recommendations in the report *Limiting thermal bridging and air leakage: robust construction details for dwellings and similar buildings*, published by the UK Stationery Office (TSO, 2001). Additionally, roofs should be designed and constructed so that the thermal transmittance (U-value) does not exceed 0.35 W/m^{2}K at any point.

A free airspace of at least 50 mm should be provided between the roof deck and the insulation. This may need to be formed using counter-battens if the joists run at right angles to the flow of air (Figure 8.18).

Where it is not possible to provide proper cross-ventilation an alternative form of roof construction should be considered.

It is possible to install vapour checks (called vapour control layers in BS 5250) at ceiling level using polythene or foil-backed plasterboard and so on to reduce the amount of moisture reaching the roof void. This is not acceptable as an alternative to ventilation unless a complete vapour barrier is installed.

Durability

According to Regulation 7 of the Building Regulations, building work must be carried out:

- A With proper and adequate materials which are
  o Appropriate for the circumstances in which they are used and
  o Adequately mixed and prepared, and applied, used or fixed so as adequately to perform the functions for which they are designed.
- In a workmanlike manner.

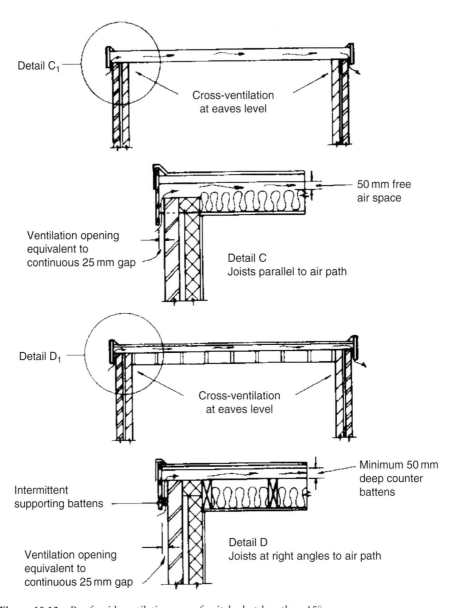

Figure 10.13 Roof void ventilation – roofs pitched at less than 15°.

Therefore, materials should be suitable in nature and quality in relation to the purpose for which, and the conditions in which, they are used.

The regulations do not specify precise durability standards in relation to materials. However, they do discuss in general terms the conditions under which a material might be considered to be unsuitable due to its short-lived nature when compared to the expected life of the building. Therefore, if a material or component is inaccessible for replacement should it fail, and its failure would create a serious health risk, then it is unlikely that the material or component would be suitable or would satisfy the building regulation standards of fitness for purpose.

Fire safety

For roofs, the most important fire safety consideration relates to the ability of the external surfaces to transmit fire across their surfaces. Fire resistance is not a factor in roof construction provided that the walls do not rely on the roof for their stability (i.e. the walls must be capable of standing independently of the roof if the roof is destroyed by fire).

The biggest risk to a roof surface in the event of fire is the possibility that the external surface may be ignited by a fire originating in a neighbouring building. If you think this is unlikely, then you should consider the fact that it was one of the main causes of the rapid spread of fire during the Great Fire of London, when the thatched roofs of many houses in close proximity became ignited and spread the fire so rapidly that it could not be extinguished. This led, in turn, to the passing of the first building legislation in England – the London Building Act – which banned the use of combustible materials on the outside of buildings in London.

The type of construction permitted for a roof depends on the use and size of the building and its distance from the boundary.

Types of construction are specified by the two-letter designations from BS 476: Part 3.

The first letter in the BS 476 designation method refers to flame penetration:

A – not penetrated within one hour;

B – penetrated in not less than half an hour;

C – penetrated in less than half an hour and

D – penetrated in preliminary flame test.

The second letter in the BS 476 designation method refers to the surface spread of flame test:

A – no spread of flame;

B – not more than 21 inches (533.4 mm) spread;

C – more than 21 inches (533.4 mm) spread and

D – those continuing to burn for five minutes after withdrawal of the test flame, or with a spread of more than 15 inches (381 mm) across the region of burning in the preliminary test.

As an example, a roof surface classified AA means that there is no fire penetration within 1 h and no spread of flame.

Table A5 to Approved Document B (an extract from which is reproduced in Table 10.2) gives a series of roof constructions together with their two-letter notional designations. In the example shown above, a roof constructed in accordance with Part 1 of Table A5 would satisfy the AA rating if it was of natural slates, fibre reinforced cement slates, clay tiles or concrete tiles and supported as shown in the table.

Table 17 of Approved Document B4 is shown in Table 10.3; it gives the notional two-letter designations for roofs in different buildings according to the distance of the roof from the boundary. Once the two-letter designation has been established a form of construction may be chosen from Table A5 of Approved Document B. Where it has been decided to use a different form of roof construction, the manufacturer's details should be consulted to confirm that the necessary designation will be achieved. It should be noted that there are no restrictions on the use of roof coverings, which are designated AA, AB or AC. Also, the boundary formed by the wall separating two semi-detached houses may be disregarded for the purposes of roof designations.

Table 10.2 Extract from Table 5 of Approved Document B – notional designations of roof coverings.

Part 1: Pitched roofs covered with slates or tiles		
Covering material	**Supporting structure**	**Designation**
1. Natural slates 2. Fibre reinforced cement slates 3. Clay tiles 4. Concrete tiles	Timber rafters with or without underfelt, sarking, boarding, woodwool slabs, compressed straw slabs, plywood, wood chipboard or fibre insulating board	AA

Part 3: Flat roofs covered with bitumen felt

A flat roof comprising of bitumen felt should (irrespective of the felt specification) be deemed to be of designation AA if the felt is laid on a deck constructed of 6 mm plywood, 12.5 mm wood chipboard, 16 mm (finished) plain edged timber boarding, compressed straw, slab, screeded wood wool slab, profiled fibre reinforced cement or steel deck (single or double skin) with or without fibre insulating board overlay, profiled aluminium deck (single or double skin) with or without fibre insulating board overlay or concrete or clay pot slab (*in situ* or precast), and has a surface finish of

a.   bitumen-bedded stone chippings covering the whole surface to a depth of at least 12.5 mm;
b.   bitumen-bedded tiles of a non-combustible material;
c.   sand and cement screed or
d.   macadam.

Notes:
∗ Lead sheet supported by timber joists and plain edged boarding should be regarded as having a BA designation.

Thatch and wood shingles

Thatch and wood shingles that cannot achieve the performance specified in BS 476: Part 3 should be regarded as having a designation of AD/BD/CD in Table 17 of Approved Document B. However, it may be possible to locate thatch-roofed buildings or extensions closer to the boundary than the distances permitted by Table 17 if the following

Table 10.3 Approved Document B, Table 17 – limitations on roof coverings.

Designation of covering of roof or part of roof	Minimum distance from any point on relevant boundary			
	Less than 6 m	At least 6 m	At least 12 m	At least 20 m
AA, AB or AC	●	●	●	●
BA, BB or BC	○	●	●	●
CA, CB or CC	○	● (1)(2)	● (1)	●
AD, BD or CD (1)	○	● (2)	●	●
DA, DB, DC or DD (1)	○	○	○	● (2)

Notes:
● Acceptable.
○ Not acceptable.
1. Not acceptable on houses in terraces of three or more houses.
2. Acceptable on buildings not listed in Note 1, if part of the roof is no more than 3 m² in area and is at least 1500 mm from any similar part, with the roof between the parts covered with a material of limited combustibility.

precautions (taken from *Thatched buildings. New properties and extensions* [the 'Dorset Model'], available at www.dorset-technical-committee.org.uk) are incorporated into the design:

- The rafters are overdrawn with construction having at least 30 minutes fire resistance.
- The guidance in Approved Document J *Combustion appliances and fuel storage systems* is followed.
- The smoke alarm system recommended in AD B1 is extended to the roof space.

FURTHER INFORMATION

References used in this chapter include the following:

- *Span Tables for Solid Timber Members in Floors, Ceilings and Floors (excluding trussed rafter roofs) for Dwellings* produced by TRADA Technology Ltd. ISBN: 1 900510 42 1

Further information can be obtained from

www.trada.co.uk
www.decra.co.uk
www.nfrc.co.uk
www.masticasphaltcouncil.co.uk
www.leadsheetassociation.org.uk
www.spra.co.uk
www.icopal.co.uk
www.dorset-technical-committee.org.uk

11 Wall and ceiling finishes

Questions addressed in this chapter:

What materials can I use for internal and external wall and ceiling finishes?
What design issues do I need to consider?
What should I consider if I need to repoint my house?
What Building Regulations apply to these construction areas?

WALL FINISHES

Irrespective of the type of construction, there are generally a variety of options to choose from when considering the finish requirements for walls both internally and externally.

The internal finish is primarily decided by the occupant's requirements/needs, whereas the external finishes can be determined by legislative requirements (e.g. Planning Permission, Conservation Area Consent, Listed Building Consent etc.).

Building Regulations can have an impact on the types of materials used on both external and internal surfaces. This is mainly to prevent or reduce the effects of fire spread should a fire break out inside or adjacent to a dwelling.

Internal finishes

If the internal surface is of masonry construction (e.g. brick, block or stone), then a number of finishes are available.

Render and set

This consists of a two-coat sand and cement mortar render (three coats for stonework) as a backing coat to eliminate any inaccuracies in the wall and provide an even surface for the two-coat gypsum plaster finish. This provides a hard, durable wall finish with a smooth surface to receive paint or wallpaper for example. This work can prove time consuming, as the individual coats of render must set before successive coats can be applied. It is also normal to incorporate a certain amount of lime into the mix to give it flexibility, but this can delay the set even further. Lime is absolutely essential where the backing is relatively weak, as in some types of stone. The design of the correct render or mortar mix is an art, especially when dealing with old buildings. Care must be taken to match the strength of the render with the strength of the backing, since, if it is too strong, it will not bond adequately and may actually damage the substrate.

Extending and Improving Your Home: An Introduction, First Edition. M.J. Billington and C. Gibbs.
© 2012 M. J. Billington and C. Gibbs. Published 2012 by Blackwell Publishing Ltd.

As stated elsewhere in this book, modern buildings tend to depend on an impervious outer layer and cavities to keep out moisture. Conversely, old buildings tend to rely on their porous nature to allow water absorbed by the fabric to evaporate back out. The use of an impervious Portland cement render in place of a traditional lime-based covering restricts evaporation. Hairline cracks form due to the mortar being more rigid than the wall. These then draw in water that becomes trapped in the fabric.

A list of reference sources on the correct mortar and render mixes for various types of backings can be found on the Institute of Historic Building Conservation's web site (www. ihbc.org.uk). Additionally, the Society for the Protection of Ancient Buildings (SPAB) provides useful information on the use of the correct render or mortar on its web site (www.spab.org.uk).

For the finish coat internally, there are a great many proprietary plasters on the market based on a range of different materials. It is usual for Portland cement, gypsum and lime to figure in these products. However, modern plasters can also be modified using polymers and other additives to improve their workability, reduce or increase their setting times, improve their flexibility and damp resistance and improve their surface resistance to abrasion and wear.

One thing can be said for certain, however, that ordinary gypsum-based plasters should never be used where there is or has been a damp problem in a wall. Gypsum will break down in continuously damp conditions and even where a wall which was previously damp and has now dried out, it will almost certainly be saturated with sulphates and other chemicals which have the ability to absorb moisture from the air (hygroscopic). Inevitably, this will cause the gypsum plaster to become wet and deteriorate. In these circumstances, sulphate-resistant cement-based plasters and renders are the best choice.

Dry lining

As an alternative to render, to save time and to eliminate a lot of the wet trade work, plasterboard can be mechanically or with adhesive fixed to the internal wall and a one- or two-coat plaster skim applied. Instead, the plasterboards could be taped and caulked ready for paint or wallpaper. Dry lining can be considerably quicker to install and considerably cheaper. This method of finish can also be combined with a thermal upgrade of the external walls (Chapter 13).

Again, special precautions should be taken to isolate ordinary gypsum plasterboard from previously damp wall surfaces. To prevent this problem altogether, there are a number of internal lining boards available which are fibre reinforced to provide a high specification for fire resistance, acoustic performance, moisture and impact resistance and racking strength.

Pointing

In a lot of refurbishment work, old brick or stone walls are often left exposed as the internal finish to maintain an aesthetic appeal to the building's historic past. If this is the case, repointing will be required to provide an even and consistent joint finish where often the old lime mortar joints are dusty and crumbling. The joints should be raked (cut) out to a minimum of 20 mm deep and repointed with a similar material to the existing bedding mortar (i.e. lime mortar if lime mortar joints exist). The finished walls are often finished with a clear sealant, such as a weak mix of PVA adhesive, to prevent dusting of the surface or are often painted with emulsion.

Internal decorative finishes

There is a great variety of internal decorative finishes available to the home owner and it is in these finishes that a person can express their own individuality. It is not the purpose of this book to cover the myriad of possible internal decorative finishes that are available; there are a great many books on interior decoration on the market. However, a few pointers can be offered to some of the pitfalls that might arise if the wrong finish is used.

Sometimes the use of a particular room will dictate the finish that is applied. This is especially true of bathrooms, shower rooms, utility rooms and kitchens. In these rooms the essential requirement is that the surface should be water resistant and easily cleaned. This results in most people choosing ceramic tiles of one kind or another, although this is not the only material that can be used. There are a number of self-coloured wall boards on the market that can provide a welcome alternative to the inevitable ceramic tiles. Clearly, it would be absolute folly to use a combustible finish such as timber boarding adjacent to a heat source such as a cooker, wood burning stove or open fire.

In fact, the Building Regulations restrict the use of untreated softwood boarding as the lining of walls and ceilings in Regulation B2. This regulation controls the amount of combustible material that can be placed on the surfaces of walls and ceilings in dwellings.

Table 1 from Section 3 of Volume 1 of Approved Document B gives the recommended flame spread classifications for the surfaces of walls and ceilings in any room or circulation space.

Different standards are set for 'small rooms', which are totally enclosed rooms with floor area of not more than $4\,\text{m}^2$, and for other rooms and circulation spaces. Small rooms are allowed to have surface linings of not lower than Class 3 and other rooms are only permitted to have linings that are not lower than Class 1 (although a slight relaxation of this is permitted to allow a surface lining of no lower than Class 3 in these other rooms provided that it does not cover an area more than half the floor area of the room).

These classes relate to the speed at which a fire would spread across the surface of the lining and would, therefore, have the ability to spread a fire rapidly throughout a room. Common ceiling materials that come within Class 1 include plasterboard and mineral fibre tiles. Common materials that come within Class 3 include timber or plywood with a density of more than $400\,\text{kg/m}^3$ whether it is painted or not. If it is desired to fully line a ceiling and the walls in a room which is larger than $4\,\text{m}^2$ in area, then proprietary clear treatments are available, which when applied to the surface of the wood will improve its surface spread of flame class to Class 1.

External finishes

As previously mentioned, the external finish can be determined by a legislative requirement, so can vary widely owing to the amount of finishes that exist in a particular location and the requirement of a building to fit in with its surroundings, adjacent properties and the existing property itself.

Render

A sand and cement mortar (or lime mortar for old buildings and conservation projects) can be applied. This can be a float finish to receive a paint finish or it can have a number of applied finishes, such as pebble or spar dash, roughcast or Tyrolean for example. Spray render applications are also available but are generally carried out by specialist contractors. This finish can also be incorporated as part of a thermal upgrade option (Chapter 13).

Tile hanging

Tile or slate hanging to vertical walls is a common finish, particularly in coastal and rural areas with quite severe exposure. It can also be used simply as a decorative feature. Timber battens are fixed vertically to the external wall and then horizontally counter-battened, to which the tiles or slates are fixed. In very exposed locations, it may be necessary to provide a breather membrane between the first line of battens and the original wall surface. The advantage of using natural materials such as slate and tile is that, if properly installed, they act as a rainscreen cladding and do not need subsequent maintenance.

Timber cladding

Timber cladding can be fixed vertically or horizontally to battens fixed to the wall. Timber has been used for centuries as an external finish and there is a variety of species available that are suitable owing to their natural resistance to decay (e.g. oak, cedar and larch). Of course, there is also a wide range of PVCu products available as cladding materials. Softwood board materials can also be used but these should have a finished thickness of at least 25 mm and should be pressure treated with preservative. They will inevitably need to be recoated with a preservative paint or microporous varnish every 3–5 years depending on the severity of their exposure.

It should be remembered that where a combustible cladding is fixed onto the outside of an existing building, it is regarded as a material alteration under the Building Regulations. This has the effect of limiting the area of the cladding depending on the distance of the relevant wall from the boundary of the site.

Table 11.1 is taken from Approved Document B, Volume 1, Dwellinghouses, and shows the maximum total of unprotected areas that can be contained in the side of a building relative to its distance from the boundary on that side. Unprotected areas include 'holes' in the wall for windows and doors, and clearly these areas are riskier from a fire safety point of

Table 11.1 Permitted unprotected areas for Method 1.

Minimum distance (*A*) between side of building and relevant boundary (m)	Maximum total area of unprotected areas (m^2)
1	5.6
2	12
3	18
4	24
5	30
6	No limit

view than combustible claddings. Therefore, it is allowed to double the areas shown for any combustible claddings that are installed provided that this is the only change made to the wall. It is necessary, however, to take into account the areas of already existing windows, doors and combustible claddings, since the rule works to the extent that an existing situation is not made worse than it is already. Therefore, the allowance is cumulative up to the maximum area.

Face brickwork

This is a self-finish wall construction whereby, once constructed, no further applied finishes need to be added. A wide variety of brick types and colours are available to achieve a range of finishes. It must be remembered that brickwork consists of bricks and mortar, and treatment of the mortar joints can have a major effect on the appearance, life and weather resistance of a wall.

For new brick walls, the strength of the bricks and mortar should be matched and, preferably, the mortar should be slightly weaker than the bricks. The reason for this is that if the building does suffer slight movement due to settlement, drying shrinkage or thermal movement, then any cracking that takes place will be in the mortar joints, where it will be distributed over a large area and will probably not be noticeable. Where any cracking is noticeable, it can be easily repaired. If the mortar is stronger than the bricks, any movement will result in the bricks cracking, which is unsightly and more expensive to repair. The mortar should also have a similar porosity to the bricks and if in a marine or other aggressive setting, it should be made with sulphate-resisting cement.

Old brick walls eventually need repointing or their ability to keep out the weather is reduced, allowing moisture penetration to the inside of the building or to materials that might be adversely affected by moisture. Mortar deteriorates because it is subject to the weather and to atmospheric pollution. Eventually, the surfaces of the joints become friable and crumble and can be removed by simply rubbing a finger along them. Over time the building will inevitably have settled and some of the joints will have opened up. When this happens, rain is able to penetrate further into the joints and in cold spells, this trapped moisture will freeze. The resulting freeze/thaw cycle will result in increasingly rapid deterioration of the mortar joints.

Repointing is done by raking out the existing mortar to a minimum depth of 20 mm. If this is skimped, the new mortar will have insufficient depth to get a good key into the joint and will fail prematurely. Good repointing should last at least 30 years and, as it is expensive to carry out, it makes sense to do it properly first time. The raked out joints should be brushed clean and then should be wetted down to prevent the new mortar from drying out too quickly, since this can cause loss of key and strength in the new mortar. The mortar mix used is of crucial importance. It should be weaker than the bricks and should be slightly porous, so that any trapped water already in the wall can evaporate out. It is normal practice to incorporate a little lime into the mortar mix and a typical repointing mix would be one part cement to one part lime to six parts sand (1:1:6). Where a weaker mortar is desired, the sand component can be increased to nine parts (1:1:9) or the cement content can be halved ($^1/_2$:1:6). Finally, the filled joints are pointed using a variety of different styles. The most common forms of pointing are bucket handle, struck and recessed (Figure 11.1).

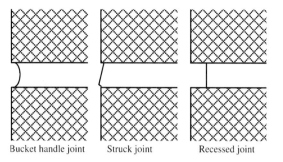

Bucket handle joint   Struck joint   Recessed joint

Figure 11.1   Common styles of pointing.

Each pointing style has its own advantages and disadvantages and it is likely that the struck joint is the best at shedding water from the wall surface. It also takes more skill to do it accurately. The bucket handle has become the norm for modern bricklayers, as it does not require great skill and is quicker to do than a struck joint. Recessed joints have the advantage of emphasising the joints in the brickwork when this is a desired feature. This style of brickwork is often coupled with a contrasting coloured mortar, which together with the shadows cast by the sharp brick edges produces a striking effect. The problem with recessing the joints is that it exposes the sharp brick edges (known as the arrises) to the weather, which allows water to lodge in the recesses. During cold weather, there is a danger that this water may freeze and eventually break down the brick edges. For these reasons, the bricks chosen must be hard, durable and frost resistant and the mortar must be equally matched in strength and porosity. Whichever joint is used, it is important in lime mortars that the joint is not 'ironed' in so that the surface of the joint becomes virtually impermeable, as this will prevent trapped moisture from evaporating through the joint surface.

CEILING FINISHES

In dwellings it is almost exclusively the case that ceilings are finished in plasterboard with either taped and filled joints or a skim coat of finish plaster. The plasterboard is usually nailed or screwed directly to the ceiling joists. In a new extension, this always leads to a degree of cracking at the perimeter or between individual board joints. This is inevitable when the boards are fixed directly to the joists, since they will tend to dry out in the warm atmosphere of the house and will, in turn, shrink and twist slightly.

Perimeter cracking can be caused by this drying movement in the joists, but it can also be caused by differential movement, where two differing structural systems meet (the joisted floor or ceiling and the brick or block walls). Perimeter cracking is usually overcome by installing a coving at the junction of wall and ceiling to mask any cracking. Cracking along board joint lines can also be caused by joist deflection in two-storey extensions where the first floor is overloaded with furniture or is subject to excessive movement of the occupants. Initial drying cracking is usually hidden when the ceiling is decorated for the second time. If, however, it is seen to recur after redecoration, then consideration should be given to reducing the load on the floor.

Although not common practice in the United Kingdom, on continental Europe it is normal practice to line the underside of the first floor or ceiling joists with substantial softwood battens or galvanised steel channel sections to which the ceiling boards are fixed and this does result in less occurrences of ceiling cracking.

FURTHER INFORMATION

References used in this chapter include the following:

■ Building Regulations, Approved Document B – Fire Safety, Volume 1, Dwellinghouses

Further information can be obtained from

www.ihbc.org.uk
www.spab.org.uk.

12 Services

Questions addressed in this chapter:

What services are covered by the Building Regulations?
What services can I install myself?
What services are best left to be installed by a competent tradesman?
What tests need to be carried out on any services that I want to have installed?
What safety precautions are applicable to the services that I want to have installed?

INTRODUCTION

Except for very minor work, such as putting up a few shelves or redecorating a room, most building projects consist of two main elements, the fabric of the building (walls, floors, roofs etc.) and the services (electricity, water supply, drainage etc.). While it is often within the capabilities of any reasonably handy (and fit) person to build foundations or walls and construct a simple roof, when it comes to services it is usually better to leave these to a competent specialist. The main reason for this concerns health and safety, because if you get the electric wiring wrong, it can kill you (or someone else) and if you get the plumbing wrong, you can flood your house. For one service in particular, the alteration or installation of a gas service or gas appliance, it is illegal for a person to do this work themselves. It must be done by a person listed on the official Gas Safe Register (www.gassaferegister.co.uk).

A number of other services are subject to control under the Building Regulations; these are known as 'controlled services'. So if you provide, extend or alter a controlled service during your project, then the Building Regulations will apply to that work just as they will to the new or altered building fabric. Additionally, most of the work to these services will not involve an application to the local authority if carried out by a 'Competent Person' (Chapter 2), since these firms are authorised to self-certify their work as complying with the regulations. While it is theoretically possible to install certain services yourself (apart from work involving gas supplies), how would you confirm that, for example, an electrical installation was safe to use? This would involve carrying out certain tests on the circuits. These tests require specialist equipment to be used and demand a high degree of training, and they would still have to be carried out by a competent person. Some services (mostly to do with above and below ground drainage) can be installed by a reasonably proficient DIY worker; this is discussed in some detail in this chapter. Others, such as the installation of boilers, heating systems, hot and cold water installations and fuel burning appliances,

Extending and Improving Your Home: An Introduction, First Edition. M.J. Billington and C. Gibbs.
© 2012 M. J. Billington and C. Gibbs. Published 2012 by Blackwell Publishing Ltd.

involve real health and safety risks and should always be carried out by suitably qualified persons.

The following services are controlled by the regulations and are concerned with the provision, extension or material alteration of

- Hot and cold water supply to bathroom fittings, sinks, hot water storage systems and sanitary conveniences (under Part G);
- Above and below ground, foul and surface water drainage and waste disposal systems (under Part H);
- Combustion appliances (under Part J);
- The provision (including replacement) or extension of the following (under Part L):
    - (i) space heating systems (and associated boilers, hot water pipes and hot air ducts etc.)
    - (ii) hot water systems (and associated boilers and hot water vessels etc.)
    - (iii) lighting systems
- Electrical installations (under Part P).

More information on the application of the Building Regulations to services installations can be found in the UK Department for Communities and Local Government publication *Domestic Building Services Compliance Guide: 2010 Edition.* This gives guidance on the following services:

- Heating and hot water systems (including insulation of pipes, ducts and vessels)
- Mechanical ventilation
- Fixed internal lighting
- Fixed external lighting
- Renewable energy systems

INSTALLING NEW KITCHENS AND BATHROOMS

While the work of refitting a kitchen or bathroom with new units and fittings does not itself generally require Building Regulations approval, the drainage or electrical works that form part of the refit may do so. Furthermore, if a bathroom or a kitchen is to be provided in a room where there was not one before, Building Regulations approval is likely to be required to ensure that the room will have adequate ventilation and drainage and meet requirements in respect of hot and cold water supply, structural stability (especially if walls are removed or a bath is installed, thereby applying an additional loading to a first floor) electrical and fire safety.

If you simply replace a WC (toilet) or washbasin, for example, and the sanitary fitting is approximately in the same place as the original, involving minimal alteration to the hot and cold water supply or drainage, then the regulations would not apply. This is termed a 'like-for-like' replacement and is treated as a repair under the regulations. Therefore, it is possible to replace an entire bathroom without needing to make a Building Regulations application provided that you do not need to install new electrical circuits or make a new connection to the below ground drainage system.

ABOVE AND BELOW GROUND DRAINAGE

Apart from carrying out necessary repairs and maintenance or doing minor alterations to above ground sanitary pipework as mentioned above, most new work to drainage systems is covered by the regulations. A competent person can self-certify work involving the installation of a sanitary convenience, sink, wash basin, bidet, fixed bath, shower or bathroom in a dwelling, provided that this does not involve work on shared or underground drainage. For work on underground drainage, an application must be made to the local authority or an Approved Inspector.

Drainage systems

These can be divided into two categories, 'foul' and 'surface water'. Foul drainage carries the soiled water from normal sanitary fittings such as toilets, basins, baths, showers and bidets and kitchen fittings such as sinks, dishwashers and washing machines. The above ground pipework for foul drainage is normally referred to as sanitary pipework, while the underground pipework is generally referred to as foul drains and foul sewers.

Surface water drainage carries rainwater, melted snow and ice from hard surfaces such as roofs and paving. The above ground disposal system of gutters and rainwater down-pipes is referred to as roof drainage, while the underground pipework is referred to as surface water drains and surface water sewers.

Drains and sewers

The distinction is made above between a drain and a sewer. In general, a drain serves a single property whereas a sewer serves more than one property. Sewers can also be described as either private or public. Private sewers are owned by the properties they serve. Public sewers are owned and maintained by the local Sewerage Undertaker. If it is intended to carry out extension or alteration work on or near a public sewer, the permission of the sewerage undertaker must be obtained. This is in addition to any compliance that may have to be obtained under planning laws or Building Regulations, although there is a duty placed on your chosen building control body to consult with the sewerage undertaker if it is thought that the works will affect a public sewer. Building over an existing drain or sewer can damage pipes, causing them to leak or become blocked, and this can lead to odour nuisance, health problems and environmental damage. It also makes it more difficult, time consuming and expensive to clear blockages and repair or replace faulty drains. So if there is an existing drain below, or close to, your proposed extension, it may need to be moved or protected, and this is likely to increase the cost of your project. Contact details for your local sewerage undertaker can be obtained from your sewerage bill.

Location of underground drains and sewers

Since it is important to know the location of underground drains and sewers, there are a number of ways in which this can done:

For public sewers, maps can be inspected free of charge at the offices of the sewerage undertaker or local authority. However, our experience is that these are not always accurate and cannot be relied upon absolutely, so caution is necessary when excavating near the assumed location of a public sewer.

Private sewers and drains are not normally mapped and their location needs to be found in other ways:

- By lifting a drain cover, it may be possible to see the direction, size and depth of pipes and by flushing toilets and so on, it may be possible to establish if the drain is foul or surface water. Deep manholes should not be entered since they can contain toxic gas.
- By studying the locations of rainwater pipes, sanitary pipework stacks and external gullies, an indication may be obtained of where their underground drain runs are likely to be.
- CCTV drain surveys can be commissioned from suitable firms. These will indicate the condition of the drains as well as their location and depth.
- The local authority may be able to provide information on the drainage runs around your house, especially if a former owner has carried out extension or alteration work and has had the work inspected by the local authority.

Separate and combined sewers and drains

In modern drainage systems, foul water and rainwater are carried separately. However, some of the older (mostly Victorian) public sewers are on a combined system taking foul and rainwater in the same pipe. Therefore, it is important to establish if your below ground drainage system is based on a separate or combined system. Most local authorities and sewerage undertakers will not allow you to connect rainwater to a foul drainage system, even if the sewers are of the old combined type. The reason is that in periods of heavy rainfall, the sewage works can be flooded and it is possible for untreated sewage to be released into rivers or into the sea, which is contrary to environmental law. The preferred way of dealing with surface water from paving or roofs is to take it to a soakaway or an existing water course. Only if all of these options are unavailable should the surface water be taken to a surface water sewer.

Connections to and removal and disconnection of below ground drains and sewers

This book is concerned with alterations and improvements to houses and, as such, does not cover the installation of totally new drainage and disposal systems. However, reconstruction and alteration to existing drains and sewers constitutes a material alteration of a controlled service under the Building Regulations and should be carried out to the same standards as new drains and sewers. Therefore, where new drainage is connected to existing pipework, the following points should be considered:

- Existing pipework should not be damaged (e.g. use proper cutting equipment when breaking into existing drain runs).
- The resulting joint should be watertight (e.g. by making use of purpose made repair couplings).

■ Care should be taken to avoid differential settlement between the existing and new pipework (e.g. by providing proper bedding of the pipework).

Even though the Building Regulations do not cover requirements for continuing maintenance or repair of drains or sewers, sewerage undertakers and local authorities have a variety of powers under other legislation to make sure that drains, sewers, cesspools, septic tanks and settlement tanks do not deteriorate to the extent that they become a risk to public health and safety. This includes powers to ensure that

■ Adequate maintenance is carried out;
■ Repairs and alterations are properly carried out and
■ Disused drains and sewers are sealed.

This is necessary because disused drains and sewers can be prejudicial to health in that they harbour rats, allow them to move between sewers and the surface, and may collapse causing possible subsidence.

Disused drains or sewers should be disconnected from the sewer system as near as possible to the point of connection. Care should be taken not to damage any pipe which is still in use and to ensure that the sewer system remains watertight. Disconnection is usually carried out by removing the pipe from a junction and placing a stopper in the branch of the junction fitting. If the connection is to a public sewer, the sewerage undertaker should be consulted.

Shallow drains or sewers (i.e. less than 1.5 m deep) in open ground should, where possible, be removed. To ensure that rats cannot gain access, other pipes should be grout filled and sealed at both ends and at any point of connection. Larger pipes (225 mm in diameter or greater) should be grout filled to prevent subsidence or damage to buildings or services in the event of collapse.

Design and construction of below ground drainage

Outfalls

The preferred way of dealing with surface water from paving or roofs is to take it to a soakaway or an existing water course. Only if all of these options are unavailable should the surface water be taken to a sewer.

On the other hand, foul water (water from bathrooms, toilets and other sanitary appliances) must always be taken to a suitable foul water outfall. In towns this will always be to a foul sewer which, in most cases, will be maintained by the local sewerage authority. In the countryside, where foul drainage may not be installed, the outfall would preferably be a wastewater treatment system or, as a last resort, a cesspool.

A cesspool is simply an impervious tank where sewage is stored until it can be removed. It is not treated in any way, so the tank must be of large capacity. Approved Document H, which deals with drainage and waste disposal, recommends that this should have a capacity of at least $18\,m^3$ based on two users. For each additional user the capacity should be increased by $6.8\,m^3$. Therefore, for a family of four, the capacity would have to be a massive $31.6\,m^3$. This is 31 600 l (or over 7000 gallons). Clearly, such a tank would be very expensive to construct and of course the owner has to pay for it to be emptied several times a year at a cost typically exceeding £150 each time.

A septic tank is really a small sewage works. The system typically includes a septic or settlement tank which provides primary treatment to the effluent from a building. The discharge from the tank can, however, still be harmful. Therefore, there is a need for a system of drainage which completes the treatment process after the effluent has passed through the tank, thus providing a means of secondary treatment.

Any septic tank and its form of secondary treatment or cesspool must be sited and constructed so that it

- Is not prejudicial to health;
- Will not contaminate any watercourse, underground water or water supply;
- Is accessible for emptying and maintenance and
- Will continue to function in the event of a power failure to a standard sufficient for the protection of health, where this is relevant (i.e. where a power supply is needed for normal operation of the system).

Furthermore, any septic tank or cesspool must be

- Adequately ventilated
- Of adequate capacity and
- Constructed to be impermeable to liquids from both outside and inside.

Since all wastewater treatment systems and cesspools rely on adequate maintenance in order to continue to operate in a safe and healthy manner, the Building Regulations require that maintenance instructions be provided in the form of a durable notice which must be affixed in a suitable place in the building. The instructions will include information on emptying at regular intervals (although this is not usually more than once a year for a septic tank and the cost will be lower than that for cesspools, since the quantity removed will be less). Traditionally, septic tanks were constructed from bricks and mortar and consisted of two adjoining tanks (an aerobic chamber and an anaerobic chamber). However, today the entire system can be purchased in a single prefabricated unit usually made from fibreglass. A typical tank is shown in Figure 12.1. The treated effluent for the septic tank will usually be fed through a filter bed and then through a series of land drains or to a watercourse; therefore, a considerable amount of land is needed to install such a system.

Below ground drainage performance requirements

The performance of a below ground foul drainage system depends on the following:

- Drainage layout
- Provision for ventilation
- Pipe cover and bedding
- Pipe sizes and gradients
- Materials used
- Provisions for clearing blockages

The drainage layout should be kept as simple as possible with pipes laid in straight lines and to even gradients. The number of access points provided should be limited to those essential for clearing blockages. If possible, changes of gradient and direction should be combined with access points, inspection chambers or manholes.

Figure 12.1   Diagrammatic section of typical modern prefabricated septic tank.

A slight curve in a length of otherwise straight pipework is permissible provided the line can still be adequately rodded. Bends should only be used in, or close to, inspection chambers and manholes or at the foot of discharge or ventilating stacks; the radius of any bend should be as large as practicable.

To give a reasonable gradient that will carry the flow and avoid blockages, the drain or sewer that you intend to connect to generally needs to be at least 0.8 m lower than the ground floor level.

Modern methods of construction rely heavily on preformed plastic inspection chambers, rodding points, access fittings and flexible plastic pipes. The only problem with such materials is that they can deform under load, so flexible pipes should be provided with a minimum depth of cover of 600 mm under gardens. Where flexible pipes have less than the minimum cover depths, they should be protected where necessary as shown in Figure 12.2.

Every part of a drainage system should be accessible for clearing blockages. The type of access point chosen and its siting and spacing will depend on the layout of the drainage system and the depth and size of the drain runs. However, for the drainage systems covered by this book, we are concerned only with rodding points and access fittings. Where there is a need to provide a deep manhole or access chamber, the services of a competent builder should be sought. For information purposes, details of typical inspection chambers in traditional construction and in modern moulded polypropylene are illustrated in Figure 12.3.

Rodding points (or eyes) are really just extensions of the drainage system to ground level where the open end of the pipe is capped with a sealing plate. Access fittings are small

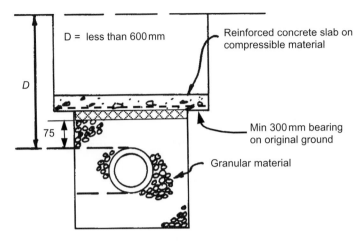

Figure 12.2   Protection shallow pipes.

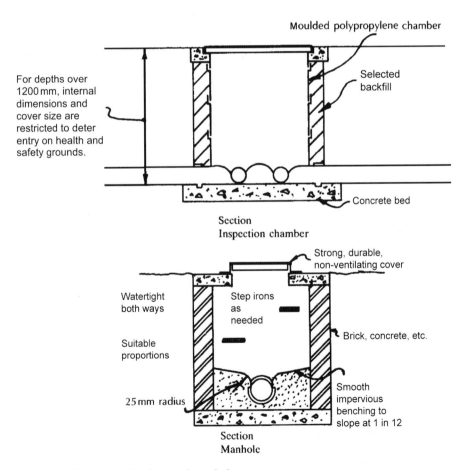

Figure 12.3   Inspection chambers and manholes.

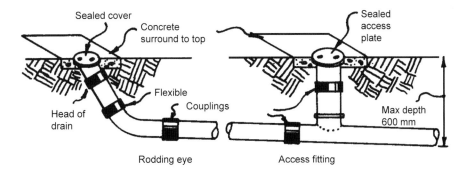

Figure 12.4 Rodding points and access fittings.

chambers situated at the invert level (bottom) of a pipe with a minimal area of open channel. Figure 12.4 shows how these two forms of access would be situated in a below ground drainage system. The pipework shown in Figure 12.4 is of the high strength stoneware type with flexible plastic joints. This form of pipework is very strong and at the same time flexible. It suffers from the disadvantage that the pipe lengths do not usually exceed 1.6–2 m in length, compared to the 6 m lengths of plastic pipes; therefore, there is a need to have many more joints. The pipes are laid in the same way as plastic pipes on a bed of small stones called pea gravel, but because of their inherent strength they do not need the protection that must be afforded to plastic pipes.

For ordinary domestic work, the maximum pipe size needed is 100 mm inside diameter; this should be laid to fall in the direction of the outlet at falls of between 1 in 40 and 1 in 80. We recommend that if high-strength stoneware pipes are used, the minimum fall should be 1 in 60, as the additional joints can mean that the flow through the pipe may be reduced due to the frictional effect of the joints.

Plastic pipe systems can be laid at falls as low as 1 in 80. However, be sure to check that the pipes have been stored on a level surface prior to use since they are prone to deformation, especially if the weather is warm. This can result in the pipe taking on a pronounced curve, which could lead to a backfall and subsequent blockage risk if the pipe is laid with the curved side down.

Surface water drainage

When building an extension to a house, it may be possible to drain the new roof area into the existing house rainwater system. However, it is often the case that either the existing rainwater gutters and downpipes need to be increased in size to take the additional flow or new gutters and downpipes need to be installed. Approved Document H contains a simple table (Table 2, reproduced here as Table 12.1) that allows gutter and downpipe sizes to be assessed depending on the roof area being created. This may result in the need to increase the size of gutters and rainwater pipes or add new rainwater pipes.

For very minor extensions such as small porches or car ports, it may be acceptable for rainwater pipes to discharge onto the ground, but you should always check with your building control body to see if this arrangement is acceptable to it. If you decide to allow rainwater pipes to discharge onto the ground, you need to make sure the water will not

Table 12.1 Approved Document H: simple method of sizing rainwater gutters and downpipes.

Max effective roof area (m²)	Gutter size (mm dia)	Outlet size (mm dia)	Flow capacity (litres/s)
6.0	—	—	—
18.0	75	50	0.38
37.0	100	63	0.78
53.0	115	63	1.11
65.0	125	75	1.37
103.0	150	89	2.16

Note: Refers to nominal half round eaves gutters laid level with outlet at one end sharp edged. Round edged outlets allow smaller downpipe sizes.

damage foundations (e.g. by encouraging it to spread out over a wide free-draining area or by providing a water butt to intercept the flow, with a suitably placed overflow which should also allow the water to spread out over a wide free-draining area). Also, you should ensure that it does not flow onto neighbouring property.

The rainwater can, of course, be taken to new or existing underground pipework, provided that this is designed to accept exclusively surface water. The final outfall can be to a suitably designed soakaway, provided that there is sufficient room on site to install it and the ground will allow the water to percolate away.

Soakaways

Soakaways should be designed to store the immediate surface water runoff and allow for its efficient infiltration into the surrounding soil. Stored water must be discharged sufficiently quickly to provide the necessary capacity to receive run-off from a subsequent rainfall event. The time taken for discharge depends upon the soakaway shape and size and the infiltration characteristics of the surrounding soil.

Soakaways serving catchment areas of less than 100 m² are usually built as square or circular pits filled with rubble or lined with dry-jointed masonry or precast perforated concrete ring units surrounded by suitable granular backfill. Although the design of soakaways should be carried out by considering storms of different durations over a 10-year period in order to determine the maximum storage volume, for small soakaways serving 25 m² or less, a design rainfall of 10 mm in 5 min can be taken to represent the worst case.

To check if the subsoil is capable of allowing sufficient percolation of water into the ground (some clays are almost impervious to water flow), a simple percolation test can be carried out as follows:

(1) Dig a 300 mm square hole at the base of proposed soakaway. For the test, this should be at least 1 m deep with the 300 square hole excavated to at least 300 mm below the bottom of the hole.
(2) Fill the 300 × 300 × 300 mm³ hole with water and allow it to seep away over night.
(3) Next day, refill the hole with water to 300 mm depth and observe the time in seconds to go from 75% full to 25% full.

Carry out the following calculation:

(a) Take the time from step 3 and divide it by 150 mm. This result will give the average time in seconds for the water to drop 1 mm $= V_p$.
(b) Repeat the test at least three times in a minimum of two trial holes.
(c) Calculate the average value of V_p.
(d) Check from the worst case scenario mentioned above (i.e. a percolation rate of 10 mm in 5 min $= 1$ mm in 30 s) if your measured percolation rate V_p can handle this. For example, if your measured percolation rate was 1 mm in 5 min (300 s), your soakaway would have to have a volume of 2.5 m^3 in order to drain an area of 25 m^2.

Rainwater can also be stored and used to flush toilets or water gardens (known as rainwater harvesting). The handling of water used in this way is subject to certain restrictions to prevent it from contaminating potable water supplies and expert advice should always be sought if it is intended to use greywater (water originating from the mains potable water supply that has been used for bathing or washing, washing dishes or laundering clothes) or harvested water. More information on greywater installations can be found in the Water Regulations Advisory Scheme Leaflet No. 09-02-04 *Reclaimed water systems. Information about installing, modifying or maintaining reclaimed water systems.*

Where it is impractical to use infiltration (e.g. because of nearby foundations, impermeable or contaminated ground or high groundwater), it is preferable to discharge it to a watercourse or, failing this, to a surface water sewer or, as a last resort, to a combined sewer (where this is permitted by the sewerage undertaker). Surface water must not be discharged into a foul drain or sewer.

Drainage from paths, patios and drives

To avoid increasing flood risk elsewhere, paths, patios and drives should be sloped towards permeable ground or be made of pervious materials. These include both porous materials (such as reinforced grass or gravel, porous concrete or porous asphalt) and permeable materials (e.g. clay bricks or concrete blocks, designed to allow water to flow through joints or voids). This approach minimises environmental impact and also avoids the cost of drainage. Surface water from hardstandings must be prevented from running onto the highway, where it could lead to accidents or cause a nuisance. Where it is impractical to drain onto pervious ground or use a pervious paving, it is preferable to keep the extra surface water on site, in order to avoid increasing flood risk elsewhere. This can be achieved by using a soakaway as described previously or some other way of allowing it to soak into the ground.

Design and construction of above ground drainage (sanitary pipework)

Sanitary pipework systems should not only convey the effluent safely to the below ground drains but they should also be ventilated to prevent the loss of trap seals, so that the air pressure in the system is kept reasonably constant. A number of terms are commonly used when talking about above ground drainage systems. A few of these are illustrated in Figure 12.5:

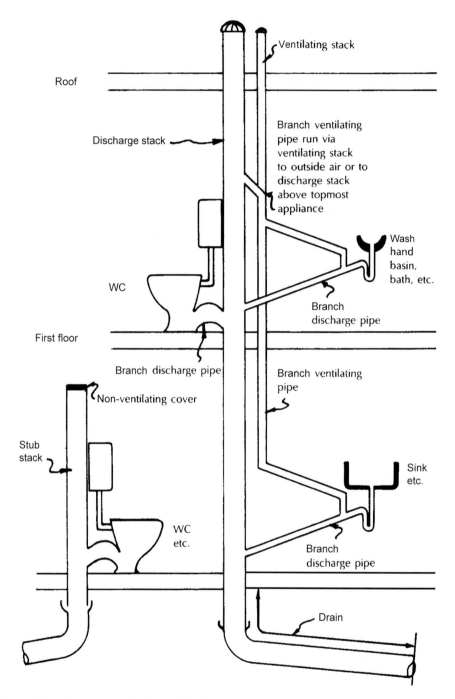

Figure 12.5 Above ground drainage definitions and principles.

Table 12.2 Flow rates for a range of appliances.

Appliance	Flow rate (l/s)
WC (9-l washdown)	2.3
Washbasin	0.6
Sink	0.9
Bath	1.1
Shower	0.1
Washing machine	0.7
Urinal (per person unit)	0.15
Spray tap basin	0.06

- **Discharge stack**: A ventilated vertical pipe which carries soil and waste water directly to a drain.
- **Ventilating stack**: A ventilated vertical pipe which ventilates a drainage system either by connection to a drain or to a discharge stack or to a branch ventilating pipe.
- **Branch discharge pipe** (sometimes referred to as a Branch pipe): The section of pipework which connects an appliance to another branch pipe or a discharge stack if above the ground floor, or to a gully, drain or discharge stack if on the ground floor.
- **Branch ventilating pipe**: The section of pipework which allows a branch discharge pipe to be separately ventilated.
- **Stub stack**: An unventilated discharge stack.

A drainage system should have sufficient capacity to carry the anticipated flow at any point. The capacity of the system, therefore, will depend on the size and gradient of the pipes whereas the flow will depend on the type, number and grouping of appliances. Table 12.2 is taken from Approved Document H, Table 2, and gives the expected flow rates for a range of appliances.

Pipe sizes and gradients

In order to work efficiently, the pipes in a sanitary pipework system need to be of certain minimum sizes and all sanitary fittings need to be separated from the main vertical discharge stack by trap seals. These trap seals are provided to prevent foul air from the system entering the building. These trap seals should remain filled with water to a depth of at least 25 mm under normal operating conditions. The gradients, pipe sizes and trap dimensions for common household sanitary and kitchen fittings are shown in Figure 12.6. The pipe sizes are nominal, as each manufacturer's sizes will vary slightly.

Ventilation of sanitary pipework

As explained, sanitary pipework needs to be ventilated to avoid foul air from the pipework and drains from escaping into the building. The normal way of doing this is to extend the vertical discharge stack to outside the building, leaving the end open (but protected with a mesh to prevent birds and vermin getting in). To stop smells entering a building, the open end of the ventilating pipe should be at least 3 m away from any opening (such as a window

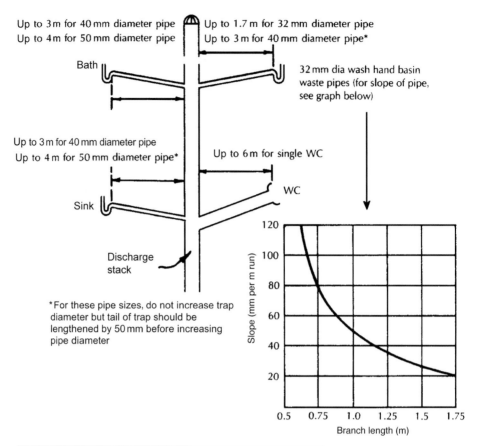

Up to 3 m for 40 mm diameter pipe
Up to 4 m for 50 mm diameter pipe

Up to 1.7 m for 32 mm diameter pipe
Up to 3 m for 40 mm diameter pipe*

Bath

32 mm dia wash hand basin
waste pipes (for slope of pipe,
see graph below)

Up to 3 m for 40 mm diameter pipe
Up to 4 m for 50 mm diameter pipe*

Up to 6 m for single WC

Sink

WC

Discharge
stack

*For these pipe sizes, do not increase trap
diameter but tail of trap should be
lengthened by 50 mm before increasing
pipe diameter

Slope (mm per m run)

Branch length (m)

Appliance	Minimum diameter of pipe and trap (mm)	Depth of trap seal	Slope (mm/m)
Sink	40	75	18–90
Bath	40	50	18–90
WC - outlet < 80 mm	75	50	18
WC - outlet > 80 mm	100	50	18
Washbasin	32	75*	See graph above

* Depth of seal may be reduced to 50 mm only with flush grated
wastes without plugs on spray tap basins

Figure 12.6   Minimum gradients, pipe sizes and trap dimensions for common household sanitary and kitchen appliances.

or skylight) into a building. If this cannot be achieved, the stack should terminate at least 0.9 m above any opening.

If the drainage is already ventilated, additional ground floor appliances (e.g. a WC and washbasin) may be connected directly to the drain without a ventilating pipe. Washbasins and sinks can also be connected to the underground drainage via a trapped gulley (Figure 12.7).

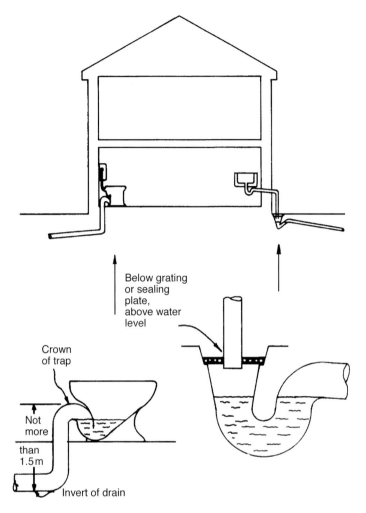

Figure 12.7 Ground floor connections for WCs and gullies.

Air admittance valves

It is permissible to terminate a discharge stack inside a building if it is fitted with an air admittance valve. This valve allows air to enter the pipe but does not allow foul air to escape. It should not adversely affect the operation of the underground drainage system, which normally relies on ventilation from the open stacks to the sanitary pipework.

Air admittance valves should also be

- Located in areas which have adequate ventilation
- Accessible for maintenance and
- Removable to give access for clearing blockages.

Air admittance valves should not be used

- In dust laden atmospheres
- Outside buildings or

- Where there is no open ventilation on a drainage system or through connected drains – other means to relieve positive pressures should be considered.

ELECTRICAL INSTALLATIONS

All electrical work that is carried out on a dwelling must comply with Part P of the Building Regulations. This does not necessarily mean that a Building Regulations application must be made for all such work. The regulations draw a distinction between what is termed 'notifiable' and 'non-notifiable' work. In Chapter 2, reference was made to Schedule 4 of the Regulations. For electrical installations, this contains, in paragraphs 1(a) to 1(e), 2 and 3, descriptions of work where no building notice or deposit of full plans is required (i.e. non-notifiable work, also called minor works).

For example, you can do the following:

- Replace any electrical fitting (e.g., socket outlet, light fitting and control switch).
- Add a fused spur (which is a socket that has a fuse and a switch that is connected to an appliance such as a heater) to an existing circuit (but not in a kitchen, bathroom or outdoors).
- Carry out any repair or maintenance work.
- Install or upgrade main or supplementary equipotential bonding (i.e. earthing of metallic objects such as copper water pipes so that an electric shock cannot be received if a fault develops in the system).
- Install cabling at extra low voltage for signalling, cabling or communication purposes (e.g. telephone cabling, cabling for fire alarm or burglar alarm systems and heating control systems).

If you are unsure as to whether or not the work you want to undertake is notifiable, your local authority building control department will advise you.

It should be noted that even minor electrical work can present a risk to safety, so you should always ensure that you know what you are doing before attempting to work on the electrical installation in your house. If a qualified electrician carries out the work, be sure to obtain a Minor Works Certificate. This will mean that the electrician has tested the work to make sure it is safe. Even if you decide to do the work yourself, it is always wise to engage a qualified electrician to check it for you.

If the work that you want to carry out is notifiable, there are a number of options open to you to obtain Building Regulations approval:

- You may use an installer who is registered with the relevant Competent Person Scheme (Chapter 2; SCHEDULE 3 – Self-certification Schemes and Exemptions from Requirement to Give Building Notice or Deposit Full Plans).
- You may make an application to your local authority building control department.
- You may use the services of an approved inspector.

The most efficient way to ensure approval is by using an electrician who is a member of a relevant Competent Person Scheme. Such a person will be able to carry out your work and self-certify to the local authority that the work is in compliance. To be a member of a Competent Person Scheme, the firm must go through a series of stringent steps to ensure that they are fully experienced and properly qualified. They must also conform to a strict code of conduct which governs the way they run their business and deal with customers. On completion of the work, the competent person should also provide you with a completed Electrical Installation Certificate, which shows that the work was tested for safety. It is also advisable to ask your electrician to provide information about which scheme they belong to and their membership number. You will then be able check with the organisation to make sure they are registered. You should also request details of the guarantee they offer on their work and check that this guarantee can be backed up with insurance in the event that they cease to trade during the guarantee period.

Installation of internal lighting systems

When new internal lighting systems are installed, they must not only be safe but also be energy efficient. This means installing energy-efficient lighting in any extension that you might add to your house or where your existing lighting system is being replaced as part of rewiring works. For example, you could install lighting fittings that have sockets that can only be used with energy-efficient lamps (i.e. those having an efficacy greater than 40 lumens per circuit-Watt).

Installation of external lights

When installing an external light (e.g. a security light) which is supplied from your home's electrical system and is fixed to the outside surface of your house, you should make sure that the light can be effectively controlled and/or that energy-efficient lamps are used. This can be achieved by

- Installing a lamp with a capacity which does not exceed 150 W per light fitting and
- Ensuring that the lighting automatically switches off at both when there is enough daylight and also when it is not required at night (e.g. by installing a lamp with a time switch, built-in light level meter and PIR infra-red motion detector).

All electrical work should follow the safety standards in BS 7671:2001, Requirements for Electrical Installations.

HEATING AND VENTILATING SERVICES

Heating and ventilating services serve two purposes:

(1) To supply heat when outside temperatures drop to the extent that comfortable living conditions cannot be maintained inside the building.
(2) To supply air thus ensuring that healthy conditions exist within the building.

In commercial and industrial buildings, such as shops, offices and factories, the heating and ventilating systems can be combined to provide air conditioning and in these circumstances, it is often the case that more energy is used cooling the air than heating it.

Air conditioning systems can be installed in dwellings but this is comparatively rare and is beyond the scope of this book.

It may come as no surprise that the Building Regulations cover both heating and ventilating systems.

Heating systems

Heating systems can be classified in a number of ways. Before the 1950s and 1960s, most houses were heated by individual fires in each room (termed local heating). These could be open fires, burning wood or coal, built on a hearth and discharging into a chimney; small gas fires that discharged the products of combustion (carbon dioxide and water vapour) into the living space or electric fires that were plugged into a normal 13 A, three-pin socket outlets and could be moved from room to room. Some electric fires were permanently wall mounted.

As living standards increased during the 1960s, it became common for houses to be built new with central heating systems and older houses were retrofitted to bring them up to the modern standards of heating.

Originally, a central heating system relied on a centrally placed source of heat, such as a gas, oil or solid fuel-fired boiler, which was connected to a series of radiators by means of pipes that carried the heat exchanging medium (water) heated in the boiler. At this time, the heat exchanger (i.e. the part of the boiler that is heated by the fuel and through which the water passes) in the boiler would have been made of cast iron. Although very robust and long lasting, cast iron heat exchangers are extremely inefficient and only convert about 50–60% of the energy derived from the fuel to the water. Such boilers have now been entirely replaced by modern high-efficiency boilers using alloy heat exchangers giving over 90% efficiency.

Other modern advances in heating technology have seen the development of ground- and air-source heat pumps and biomass boilers. Such systems are complex and expensive to install and demand a great deal of specialist knowledge if they are to really save money for the householder. Additionally, they are not really suitable for the type of work covered by this book. Nevertheless, some sources of information for these are given in Chapter 13.

Today, a modern house will probably have a fully controlled, energy-efficient central heating system coupled with an efficient wood or multi-fuel burning stove in the principal living room to give a focal point to the room and a sense of well-being that such real-flame appliances are said to give.

It is usual for the central heating system to heat the water for the house for use in kitchens and bathrooms. Originally this would be done by storing the heated hot water in an insulated cylinder. The water could then be drawn from this cylinder when needed and the cylinder would be topped up from the house cold water supply, meaning that this cold water would then have to be heated to the right temperature. Such a system has two major disadvantages. The stored water will slowly cool down when it is not being used and will need to be reheated at intervals depending on the setting of the system controls. If a large quantity of hot water is needed (such as if someone in the house has a large bath!), this will result in a delay while the stored water is again reheated.

It is estimated that over 50% of new heating and hot water installations fitted in houses today use combi (i.e. combination) boilers. A combi boiler is both a high-efficiency water heater and a central heating boiler, combined within one compact unit. Therefore, no separate hot water cylinder is required, which means that space can be saved within the property. The downside is that whereas the hot water cylinder would usually be mounted in the airing cupboard and would supply a continuous low level of heat to air clothes, without the cylinder this facility does not exist. Most people install a small radiator in the airing cupboard to make up for this, but this will be turned on and off at intervals by the heating system controls, so will not give continuous heat. This is unimportant if clothes are properly dried and aired before being placed in the airing cupboard, as the intermittent heating will usually be sufficient to keep the clothes sufficiently dry.

Further benefits of a combi boiler are significant savings on hot water costs and the fact that hot water is delivered through the taps or shower at mains pressure, enabling showering to take place without the need for a pump. In such a system, it is most important that the shower is thermostatically controlled to safeguard against sudden changes in water temperature. Additionally, a combi boiler will generally save on installation time and costs, since there will be no feed tank in the roof space and, therefore, less pipework will be needed and the installation time will be shorter and less disruptive.

Replacing a boiler

The Building Regulations will apply only if you decide to change your existing hot water/central heating boiler or if you decide to change to one of these boilers from another form of heating system.

Work to install a new boiler (or a cooker that also supplies central heating such as an Aga, Rayburn etc.) needs Building Regulations approval because of the safety issues and the need for energy efficiency. This is generally achieved by employing an installer registered under a relevant Competent Person Scheme (Chapter 2; SCHEDULE 3–Self-certification Schemes and Exemptions from Requirement to Give Building Notice or Deposit Full Plans). The competent person must follow the guidelines set out in Approved Document J, which shows what is necessary for air supplies, hearths, flue linings and chimney labelling, and where the flue outlet can be positioned. Approved Document J is complex and the safety issues covered are particularly important, since incorrect installation of a fuel burning appliance can have extremely serious consequences leading to

- The potential build-up of poisonous or suffocating gases in the building unless an adequate supply of combustion air is provided
- The entry of noxious fumes into the building from outside unless the products of combustion are discharged safely to outside air and
- Damage by heat or fire to the fabric of the building from heat produced by the appliance.

The various types of fuel burning appliances covered by the regulations are illustrated in Figure 12.8 and the principles of their operation are shown. The figure distinguishes between open-flue appliances (e.g. the traditional open fire or stove which draws its combustion air from the room or space in which it is installed and which requires a flue to discharge its products of combustion to the outside air) and room-sealed appliances (an appliance that obtains combustion air either from a ventilated uninhabited space within the building or directly from the open air outside the building. The products of combustion will be vented directly to open air outside the building).

You should not attempt to install a fuel burning appliance unless you are suitably qualified and experienced, since the potential for serious injury or death resulting from

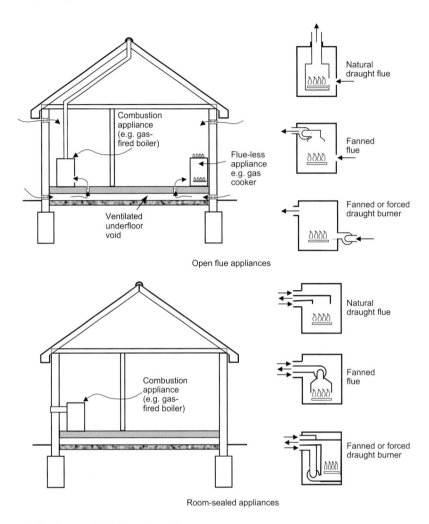

Figure 12.8 Types of fuel burning appliances.

incorrect installation is too great. You are not allowed to alter or install a gas burning appliance unless you are listed on the official Gas Safe Register.

If emergency works are necessary (because, for instance, a hot water cylinder springs a leak), there is no bar on carrying out repairs straightaway but the repair works must comply with the requirements and after the event it is necessary to apply for retrospective approval and a completion certificate.

Flues, chimneys, fuel tanks and so on

The Building Regulations also apply to the following works in connection with heating services:

- Installation of new flues for fuel burning appliances
- Construction of new chimneys
- Installation or replacement of oil and liquid petroleum gas storage tanks

This is because it is necessary to take into account such factors as ventilation, discharge of combustion products, structural stability, fire safety, environmental protection and general safety. Installation should always be carried out by a suitably qualified installer.

For fuel storage tanks, if the installation is above ground, the requirements will be applied to achieve adequate shielding of the tank from any surrounding fire and, in the case of an oil tank, containment of oil leakages so that ground water is not contaminated. Where new oil connecting pipework is proposed, a fire valve will be needed at the point where the pipe enters the building.

If you are installing an oil tank and/or connecting pipework and you employ an installer registered with one of the related Competent Person Schemes, you will not need to involve a Building Control Service.

Ventilating systems

The general purpose of ventilation in a dwelling is to remove contaminated indoor air and replace it with fresh outdoor air. We make the assumption (as does Part F of the Building Regulations) that outdoor air is relatively uncontaminated, of reasonable quality and with low levels of pollution. The following are the main purposes of ventilation:

- Removal of carbon dioxide (CO_2) and the provision of fresh air for the occupants to breathe.
- Dilution and removal of airborne pollutants, including odours. This includes not only pollutants produced by normal human activity but also those released from the materials and products used in construction, decoration and furnishings.
- Removal of excess moisture in order to control the humidity of the indoor air.
- Removal of excess heat in order to control indoor temperatures.
- Provision of air for fuel burning appliances.

Ventilation is covered by three different parts of the Building Regulations depending on which aspects are being considered:

- Guidance on fresh air ventilation is given in Approved Document F.
- Ventilation air for combustion appliances is dealt with in Approved Document J.
- Ventilation to remove excess heat is covered in Approved Document L mainly for non-domestic buildings.

In this section we are concerned with fresh air ventilation for dwellings covered by Approved Document F in extensions, renovated bathrooms and kitchens.

When you build an extension you must make sure that you provide what is termed 'purge ventilation' and 'background ventilation'. If the extension contains a bathroom, kitchen or utility room, you will also have to provide 'intermittent extract ventilation'.

Purge ventilation is also called rapid ventilation and is designed to rapidly dilute high concentrations of pollutants, such as when you carry out occasional activities like painting and decorating (or burn the toast!). This is usually catered for by having a window that you can throw open. For normal opening windows that open more than 30° and for sliding sash windows, the opening light area needs to be at least 1/20th of the floor area of the room in which it is situated. Where the window opens only between 15° and 30°, its opening area has to be at least 1/10th of the floor area of the room.

Table 12.3 Minimum extract ventilation rates (intermittent extract).

Room	Extract rate (l/s)
Kitchen, fan in extract hood over or adjacent to hob	30
Kitchen, elsewhere and not in extract hood over or adjacent to hob	60
Utility room	30
Bathroom (or shower room)	15
Sanitary accommodation such as separate WC	6

Background ventilation is usually dealt with by providing trickle ventilators in the tops of window and door frames. In this case, the ventilation area provided needs to be at least 5000 mm^2. This sounds quite a lot, but is in fact only equivalent to an opening that is 50×100 mm^2 (about 2×4 in^2). Most people encounter trickle ventilators when they have their windows replaced for the first time. The regulations say that if the windows being replaced have trickle ventilators, then the replacement windows must have them too. Conversely, if the removed windows do not have trickle ventilators, then there is no need for them in the replacements. Our advice would be that if you are having your windows replaced irrespective of whether or not the old windows had trickle ventilators, always insist on having trickle ventilators in your new windows. The simple reason for this is that modern windows are completely airtight when closed and by not having the trickle ventilator option, you may be storing up problems in terms of condensation and mould growth in your house.

Intermittent extract ventilation needs to be provided by a mechanical extract fan that draws air from the room and exhausts it to the outside air via a short duct. The extract rate of the fan needed will vary according to the type of room and the position of the fan in the room as shown in Table 12.3.

Mechanical intermittent extract ventilation needs to be adequately controlled and this can be done manually, automatically or by a combination of these methods. Where the extract ventilation is provided in a room without a window that opens (such as a bathroom or WC), the extract fan should be connected to the light switch in such a way as to continue to run for at least 15 min after the light is switched off. Many people when encountering such a system for the first time assume that the extract fan is faulty when it does not turn off with the light switch. It is not faulty and under no circumstances should it be disconnected. If this is done, it will lead to a build-up of condensation on wall and ceiling surfaces, which, in turn, will lead to the growth of black spot mould that is unsightly and destroys wall finishes.

HOT AND COLD WATER SUPPLY

Cold water supply

It goes without saying that if you install new appliances that use water (kitchen, bathroom and WC fittings and water for heating systems), then that water has to be wholesome and uncontaminated. In most cases, the water supplied to the house will be derived from the mains supply and this will be water collected and treated by your local water supply company. If a building is supplied with water from a source other than a statutory water undertaker or a licensed water supplier, then the water is considered wholesome if it meets the criteria set out in the Private Water Supplies Regulations 2009 (SI 2009/3101) for a building in England or the Private Water Supplies (Wales) Regulations 2010 (SI 2010/66) for a building in Wales.

If some or all of a wholesome water supply is softened and after softening it still meets the criteria for wholesome water, it is called softened wholesome water and may be considered wholesome and suitable for drinking. However, some softening processes result in the water having a high sodium content. If after softening the water meets all the requirements for wholesomeness except for its sodium content, it is called wholesome softened water and it should not be provided to any draw-off point where it may be used for drinking or food preparation.

It is possible to use water from sources other than the mains supply for some uses in a dwelling. Alternative sources of water include

■ Water abstracted from wells, springs, boreholes or water courses;
■ Harvested rainwater and
■ Reclaimed greywater.

Water from these sources, even after treatment, is unlikely to reach the standards required for wholesome water and should not be used for drinking or food preparation. Nevertheless, such water can be used in applications where wholesome water is not necessary. Possible applications include

■ Flushing of toilets;
■ Washing machines;
■ Washing of exterior surfaces and cars and
■ Irrigation of gardens and allotments.

The distribution of water to appliances in the building is covered by Part G of the Building Regulations and guidance is provided in Approved Document G concerning

■ Cold water supply;
■ Water efficiency;
■ Hot water supply and systems;
■ Sanitary conveniences and washing facilities;
■ Bathrooms and
■ Kitchens and food preparation areas.

It is worth noting that if you fit a washbasin adjacent to a toilet, it must be separated (by a door) from a place where food is washed or prepared and the basin itself must not be used for washing or preparing food.

In today's world with our ever-increasing population and modern lifestyles, we have an ever-increasing need for water in our homes. In light of the water shortages that have occurred since the 1970s, the Building Regulations have changed in recent years to require that we use our limited supplies of water more efficiently. The efficiency requirements of the regulations do not in fact apply to alterations and extensions (only to new dwellings or dwellings created when a building's use is changed to provide additional dwellings), so although we do not have to worry about this issue, it is good practice to install fittings and fixed appliances that use water efficiently to prevent undue consumption of water. Your builder should be able to advise you on appliances that use less water.

Hot water services

Under the regulations, hot water has to be supplied to the following appliances:

- Any washbasin provided in association with a sanitary convenience (whether in the same room or not).
- Any washbasin, bidet, fixed bath or shower in a bathroom in a dwelling.
- Any sink in a food preparation area

There is a variety of ways of heating the water for supply to the sanitary fittings and sinks listed; some of these have been considered under the heating systems section. The most important factor in any system that heats water is that it should do this safely. Water heated by a combi boiler must be delivered at the right temperature, so that the users are not going to get scalded when they wash or shower. Some hot water installations incorporate a storage vessel for hot water as already described. In a traditional system, this storage vessel is ventilated to the atmosphere and incorporates a feed and expansion tank usually mounted in the roof space of the dwelling. This serves three main functions: it provides a hydrostatic head for the appliances, so that the water comes out at reasonable pressure; it acts to top up the hot water and heating system to replace water lost by evaporation; it can accept the expansion of water in the system, which inevitably happens when water is heated.

The main problems with such a system are that the tank in the roof is costly to install, it imposes a loading on the roof structure that has to be allowed for, it is a continuing maintenance problem and it is not a suitable system for flats where a roof space may not be available. Where a storage vessel is desired, it is possible to install an unvented one provided that a series of safety precautions is taken. In such a system, the stored water is heated in a closed vessel. Without adequate safety devices, an uncontrolled heat input (caused by a malfunction of the heating system) would cause the water temperature to rise above the boiling point of water at atmospheric pressure (100°C). At the same time, the pressure would increase until the vessel burst. This would result in an almost instantaneous conversion of water to steam with a large increase in volume producing a steam explosion. Therefore, unvented hot water systems must incorporate devices designed to

- Prevent the temperature of the water stored in the vessel at any time exceeding 100°C and
- Ensure that any discharge from the safety devices is safely conveyed to where it is visible but will not cause a danger to persons in or about the building.

The regulation requirements for each system are described below. A vented system must have a

- Vent pipe;
- Safety device which is able to disconnect the supply of heat to the storage vessel
- Safety device to safely discharge the water in the event of significant overheating and
- Cold water storage cistern into which the vent pipe discharges.

Unvented systems must be protected against the build-up of excessive pressure. The protection must include the following as a minimum:

- A thermostat to control the temperature of the stored water.
- At least two independent safety devices.

A typical solution would be to provide, in addition to the water temperature thermostat,

- A non-self-resetting energy cut-out which will disconnect the supply of heat in the event of system over-heating and
- A temperature relief valve (or a combined temperature and pressure relief valve) which, in the event of serious over-heating, will discharge water safely in a place where it can be seen.

These unvented systems come as ready-built packages with all their safety devices incorporated. Therefore, they are not suitable for DIY installation.

FURTHER INFORMATION

References used in this chapter include the following:

- Building Regulations, Approved Document F – Ventilation
- Building Regulations, Approved Document G – Hygiene
- Building Regulations, Approved Document H – Drainage and Waste Disposal
- Building Regulations, Approved Document J – Combustion Appliances and Fuel Storage Systems
- Building Regulations, Approved Document L – Conservation of Fuel and Power
- Building Regulations, Approved Document P – Electrical Safety

Further information can be obtained from the following:

- For 'greywater' installations – Water Regulations Advisory Scheme Leaflet No. 09-02-04, *Reclaimed water systems: Information about installing, modifying or maintaining reclaimed water systems.*
- BS 7671:2001, Requirements for Electrical Installations.
- Department for Communities and Local Government publication *Domestic Building Services Compliance Guide 2010 Edition.*
- Private Water Supplies Regulations 2009 (SI 2009/3101) for a building in England or the Private Water Supplies (Wales) Regulations 2010 (SI 2010/66) for buildings in Wales.
- www.worcester-bosch.co.uk
- www.planningportal.gov.uk
- www.gassaferegister.co.uk

13 Improving the thermal efficiency of the home

Questions addressed in this chapter:

How can I make my home more energy efficient?
What materials can I use for insulating my home?
What design issues do I need to consider?
Are there any new materials and methods in use for these constructions?
What grants or other initiatives are available for me to improve the thermal efficiency of
  my home?
What Building Regulations apply to these construction areas?

THERMAL EFFICIENCY

Improving the thermal efficiency of the home has been a constant theme for several decades. Amidst continual legislation changes and the later emphasis on reducing carbon emissions, improving the home in this way also serves to produce a more controllable and comfortable internal environment as well as saving money for the occupants.

Heat losses from a property are basically wasted money. It is warm air that you have paid to heat via your heating system. Therefore, the more you can restrict the flow of heat the more money you can save.

When considering improving the thermal efficiency of the home we are concerned with the elements which make up the external envelope. Each of these elements can be upgraded or improved to reduce heat losses, these elements are

- Roof,
- External walls,
- Windows and doors and
- Ground floor.

When considering the thermal performance of a building element, its ability to resist the flow of heat in terms of a U-value (measured in W/m^2K) is measured. A U-value (or thermal transmittance coefficient) takes into consideration the thickness and thermal conductivity (k value or λ value) of each material making up that element. The thermal conductivity of a material is an individual material property that is very important as it

Extending and Improving Your Home: An Introduction, First Edition. M.J. Billington and C. Gibbs.
© 2012 M. J. Billington and C. Gibbs. Published 2012 by Blackwell Publishing Ltd.

refers to the amount of heat the material will conduct and, therefore, the lower the value the better. This thermal conductivity, together with the thickness of the material, will give the material resistance to the passage of heat.

A U-value is the reciprocal of the sum of all the resistances (r values) making up the element. Therefore, the higher the value of these resistances, the greater is the resistance to the passage of heat. Since the U-value is the reciprocal of the resistances, it is a measure of the transmission of heat (rather than resistance), so the lower the U-value the better, as it shows that less heat will flow.

See also Case Study 1 – Thermal Upgrading.

Roof

Insulating the roof space is a common and very cost effective means of reducing heat losses. For some people there may even be grants available to get this work done (Chapter 3). The insulation material choice largely depends on the location of the insulation within the roof space.

Insulating at ceiling joist level is the more common and most economic form of improving the roof space. Any insulation that exists can be added to with additional layers of insulation materials, such as mineral wool. Insulating at ceiling joist level consists of two layers. The first layer of 100–150 mm mineral quilt is laid between the ceiling joists, the second 150 mm layer is laid at right angles on the top the first layer so as to cover the joists (Figure 13.1). This reduces the thermal bridging effect of the joists as these are able to transmit heat at a much greater rate than the insulation.

It is important to wear appropriate personal protective equipment (PPE) when using insulation products to prevent inhalation of any fibres and prevent any skin irritation and rashes.

As the insulation is at ceiling joist level, a cold space is created above it in the loft space and a cold surface exists at rafter level. As heat inevitably rises and heat loss takes place in the roof space, any moisture passing through into this space could potentially condense on

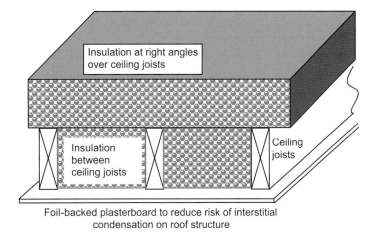

Figure 13.1   Insulating at ceiling joist level.

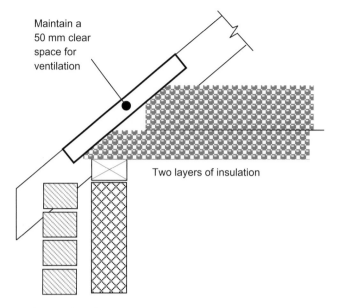

Maintain a
50 mm clear
space for
ventilation

Two layers of insulation

Figure 13.2   Ventilation of roof space.

the roofing underfelt and rafters. It is vital, therefore, to ensure that sufficient ventilation is provided within the roof space to remove any moisture-laden air, so reducing the risk of condensation and subsequent defects such as rotting timbers.

With this in mind, it is essential, when laying additional insulation between the ceiling joists, that any ventilation provided at eaves level is not blocked by the new insulation and a clear air space can be maintained on both sides of the building between the underfelt and the insulation to create cross ventilation (Figure 13.2). Proprietary rafter ventilation trays are available to ensure this air space can be maintained.

Insulation can also be provided at rafter level. The choice of insulation for this purpose would normally be a rigid slab insulation, which can be cut to the appropriate width and inserted between the rafters. It is still important to maintain a clear ventilation space between the insulation and the underfelt. However, the risk of a defect caused by condensation is reduced by putting the insulation in this location (Figure 13.3). This will improve the thermal performance of the roof to a degree dependent on the depth of rafter and the thickness of insulation to maintain a ventilation space. Therefore, to achieve the required level for Building Regulations if carrying out a loft conversion, often another deeper rafter will have to be placed (and fixed) beside the existing rafters to give more depth to receive more insulation (see Case Study 2 – Loft Conversion).

Advances in modern material technology have provided multifoil layer insulation products (for example Thinsulex and Tri Iso) that can achieve high levels of thermal insulation (U-values) at a significant decrease in thickness, some as thin as approximately 30 mm. These materials can be expensive in comparison to other forms of insulation but the savings in thickness may outweigh the cost. These materials can prove very useful in this respect when carrying out loft conversions where maximising of any headroom is essential (Figure 13.4).

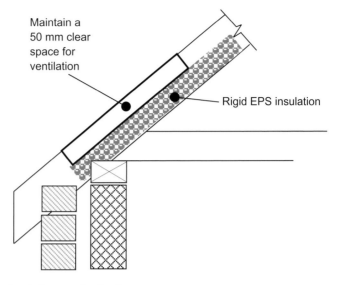

Figure 13.3 Insulating at rafter level.

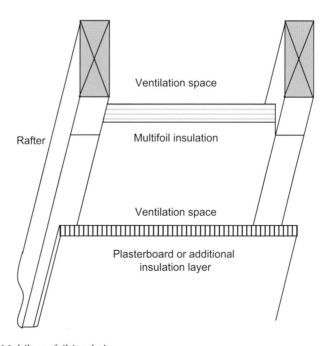

Figure 13.4 Multilayer foil insulation.

External walls

External walls can have additional insulation applied to them in three locations:

- Externally.
- Internally.
- Cavity (if one exists).

External insulation can be applied to the external face of the wall by mechanically fixing or adhesive fixing slab insulation and applying a render finish over the top. Whilst traditional renders can be applied over the insulation, proprietary thin coat polymer modified renders are more usual and reliable. These renders usually come as a complete system from manufacturers such as Sto Ltd (www.sto.co.uk) or Wetherby Building Systems Ltd (www.wbs-ltd.co.uk), and since serious problems can be caused by incorrect installation it is always better to use a recognised system fitted by an accredited installer. It will also be possible to insure the system for up to 20 years against failure when using reputable product systems. The render systems consist of the insulation (EPS, Rockwool or glass fibre batts), a base coat reinforced with a glassfibre or polyester reinforcing mat and a thin topcoat which can be obtained in a variety of colours to reduce the need for continual repainting of the external walls (Figure 13.5). Of vital importance in installing such systems is the need for adequate preparation of the external wall surface to receive the insulation. This should be cleaned to remove any mould or plant growths, such as ivy and lichen which might cause problems when enclosed by the render system. The surface to receive the insulation boards must also be made reasonably level, so that a good key can be

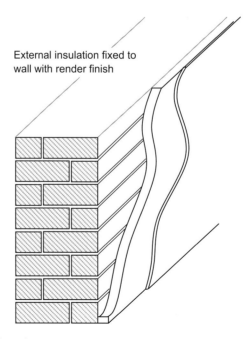

External insulation fixed to wall with render finish

Figure 13.5 External insulation.

obtained for the insulation boards or batts. Some external wall materials, such as Bradstone, have quite heavily indented surfaces and would probably need to be flushed up with a suitable levelling coat before application of the external wall insulation system.

Care must be taken when applying external insulation. Insulation must also be applied to the reveals, although this is generally thinner than the insulation on the external wall face, so as not encroach on the door/window frames and glazing.

The insulation can be in excess of 50 mm and this can cause a problem at cill level, as the cill will not be able to perform one of its functions by not being able to shed rainwater away from the building. Poor detailing or poorly applied insulation at cill level can allow water to transfer to the inner leaf. To prevent this, a plastic or powder coated aluminium subcill can be inserted below the existing cill so that rainwater can be ejected away from the wall (Figure 13.6). It is also possible to fix a cill extension over the top of the existing cill, which has the advantage that the insulation can be continuous and not give rise to a possible thermal bridge.

As external rendering changes the external appearance of a building and as the face of the building will now project from the face of any adjacent properties, planning permission may be required on standard properties but will definitely be required on buildings in a conservation area or if the building is listed.

External insulation causes minimum disruption to the occupants of the home as no work needs to be carried out internally. The cost of externally insulating a building can be offset if the building is in need of a new external render finish, as this would be included in the works and it may be possible to carry out the works without removing the existing render if this has a sufficient key to the wall, although it would probably need local repairs and removal of loose patches. As mentioned above, external insulation applications

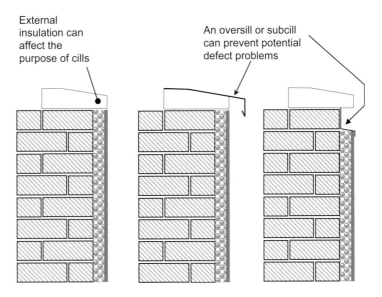

Figure 13.6   External insulation cill details.

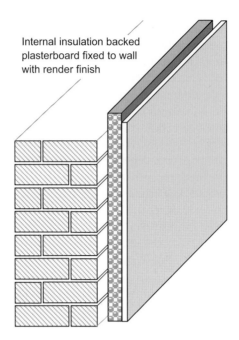

Internal insulation backed
plasterboard fixed to wall
with render finish

Figure 13.7   Internal insulation.

should only be carried out by a specialist contractor, which will also provide a guarantee for the work carried out and can be insurance backed if the client wants this.

Internal insulation is a common means of improving the thermal efficiency of an external wall and is commonly carried out by applying polystyrene-backed plasterboard to the wall. These boards can be fixed either mechanically or by the use of adhesives and can then receive a gypsum plaster skim when finished (Figure 13.7).

At first glance internal insulation may appear very cost effective and quite simple to carry out. However, there are additional items that need to be considered that could increase the cost significantly, such as

- Electrical work needed to adjust the level of socket outlets, light switches, telephone and aerial sockets;
- New skirting boards;
- Total redecoration of the rooms affected;
- Repositioning of radiators and possible redirection of services pipework and
- In kitchens, possible resiting of cupboards and fixed electrical appliances.

The above list also identifies that there will be a significant disruption to the occupants of the house and, of course, the internal dimensions of the rooms will be reduced by twice the thickness of the insulation system fitted. This can have an effect on the furniture layout in the affected rooms.

There are also a number of technical problems that internal insulation can give rise to, such as

- The need to ensure that if the building is affected by rising damp due to a faulty or non-existent horizontal damp-proof course (dpc) then this must be cured before the application of the insulation system or the damp will be trapped behind the insulation and will cause untold damage;
- The need to ensure that the walls are repaired externally and do not allow dampness due to rain or snow to penetrate, or the same problems that are associated with faulty dpc may occur;
- Placing insulation on the inside face of a solid brick or block wall can give rise to a phenomenon known as interstitial condensation, whereby moist air generated inside the building passes through the insulation layer and condenses either at the interface with the brickwork or within the brickwork itself. This can cause failure of the bond between the insulation and the wall and can give rise to mould growth within the system. This can be largely solved by incorporating a vapour control layer (VCL) between the plasterboard face and the insulation, although the work has to be done extremely carefully so that all possible penetrations of the VCL are sealed (e.g. for electric sockets etc.).

If the external walls are of cavity wall construction, then an alternative to either external or internal insulation is to have the cavities filled with insulation material. This is carried out by specialist contractors who will provide a guarantee for the work which can be insurance backed. This is done by drilling a series of holes in the wall (externally or internally) and pumping the insulation material into the holes before plugging them with mortar (Figure 13.8).

As the cavity was originally designed as an air space to prevent the passage of moisture, care should be taken in the appropriate choice of material used for the insulation and advice should be sought if the property is in a particularly exposed location to ensure that lateral penetration of moisture through the wall will not occur.

A benefit of cavity wall insulation is that there is little disruption to the occupants of the house and there should be no need to redecorate. There are also government grants available, whereby homeowners can obtain subsidised, and sometimes free, cavity wall insulation (Chapter 3).

Windows and doors

Compared with other building elements, windows and doors generally offer the least resistance to the passage of heat, so have high heat losses. Even by current standards the U-value for a window is approximately 6 times greater than that for a wall and 10 times greater than that for a roof. This information itself explains one reason why limiting the amount of glazing in a building reduces heat losses.

Of course, modern materials and methods have allowed us to reduce the impact of the rate of heat losses, but if we want to be able to see the outside world, and allow a reasonable amount of daylight into our houses (which, curiously enough, saves energy by reducing our reliance on artificial lighting) there is only so much that can be done whilst keeping costs to a reasonable proportion in relation to the rest of the build.

For example, a single glazed window would have a U-value of about $5.4\,W/m^2K$ but a modern replacement double glazed window would have to be as low as $1.6\,W/m^2K$ to satisfy current Building Regulations. Many windows are available today with U-values as

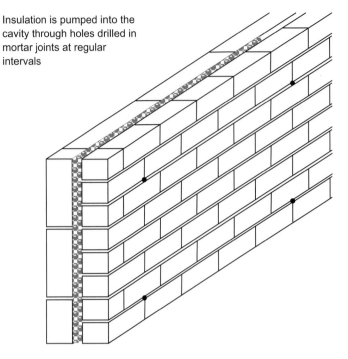

Insulation is pumped into the cavity through holes drilled in mortar joints at regular intervals

Figure 13.8   Cavity wall insulation.

low as $1.0\,\text{W/m}^2\text{K}$. It is clear to see that making this improvement would considerably reduce the heat losses through the glazing.

Current Building Regulations require new and replacement windows either to be U-value rated (the maximum permissible U-value is $1.6\,\text{W/m}^2\text{K}$) or they can be energy rated. Such windows will carry an energy label similar to that which electrical white goods (refrigerator, freezer, cooker etc.) have displayed for many years. This label shows the window's performance on an A–G scale, with A being the most efficient in terms of U-value and hence heat losses (Figure 13.9). Building Regulations require that the energy rating of a window must be of band C or better (i.e. A, B or C) and there are plans to increase this incrementally over the coming years in line with the UK Government's commitment to battle global warming.

Ground floor

As ground floors are mainly of two types (ground bearing and suspended), there are different options for improving the thermal performance of this particular element.

Ground bearing floors

New ground bearing floors have insulation built in as part of the construction to comply with current Building Regulations (Chapter 7). However, improving the insulation of an

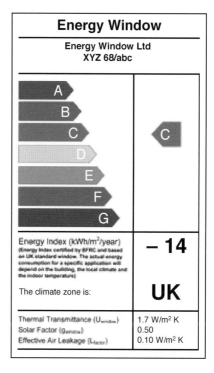

Figure 13.9 Window energy label.

existing ground bearing floor can prove more problematic, as it would increase the thickness of the floor and, therefore, its height. This has a knock-on effect on internal joinery items such as skirting boards, architraves, door linings, doors and staircases, which would greatly increase the cost of such work.

Therefore, the most feasible option to improve the insulation within an existing ground bearing floor would be to remove the sand/cement screed, thereby leaving the oversite concrete slab exposed. This screed is normally approximately 50 mm thick. As a result of this, 25 mm timber battens could be secured to the oversite concrete slab with 25 mm rigid insulation between. Timber floorboards or tongue-and-groove sheet particleboard could then be used as the new floor covering (Figure 13.10).

It is of course essential when doing this to ensure that the existing concrete slab is absolutely dry and has a fully functioning damp-proof membrane (dpm) incorporated into it. Floors can be tested with a moisture meter and if there is any sign of residual dampness then it will be necessary to install a new dpm. This can be achieved by applying a liquid membrane (such as three coats of bituminous emulsion or 10 mm of asphalt) or a sheet membrane such as 1200 gauge polythene. The battens used should be treated with preservative and if a particle board is used it should be water resistant.

The insulation used should be flooring grade. There are also some products available where the flooring grade insulation is bonded directly to water resistant particleboard, thus removing the need for battens and speeding up the whole process.

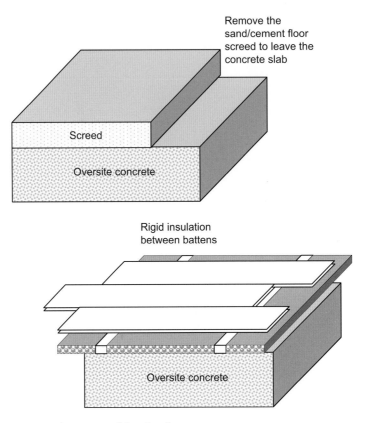

Figure 13.10 Improving a ground bearing floor.

Suspended timber floor

These floors have the scope to produce a greater insulation improvement than ground bearing floors. As there is a void beneath the floorboards, insulation can be installed between the joists. This would generally be in the form of rigid insulation (EPS, PUR, XPS etc.) resting on battens fixed to the side of the joists (Figure 13.11).

Alternatively, a quilt insulation can be installed between the joists resting on a net fixed to the underside of the joists (Figure 13.12).

The existing floorboards will clearly have to be removed and re-fitted in order for any such work to be done. Therefore, this is labour intensive work rather than expensive material costs. This also has the advantage of reinstating existing materials and maintaining the existing floor height. Care should be taken when carrying out this work that any underfloor ventilation systems are not impeded.

Beam and block floors

If these floors have been finished with a sand/cement screed then they can be treated the same as ground bearing slabs. However, if they have been finished with a tongue-

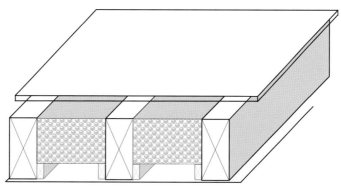

Rigid insulation laid between floor joists supported on battens

Figure 13.11 Improving a suspended timber floor.

and-groove particleboard (or similar) then this would be placed directly on the floor and no scope would exist to install additional insulation without increasing the floor height.

Since these floors are relatively new (approximately 25 years) then the blocks used between the beams may be of a lightweight variety and so may afford a reasonable level of insulation, thereby negating any expensive remedial work for little gain.

Internal environment

The positioning of the insulation in an individual element is important in relation to the control you will have over the internal environment. The further away from the internal environment the insulation is placed, the less control you will have. This is due to the element becoming a thermal store during the heating of the room, therefore taking longer to heat the room and returning some of that heat back into the room when the heating is turned off (Figures 13.13 and 13.14).

This may sound detrimental, However, it is a personal choice whether you would want that level of control. Some people prefer the building to act as a thermal store and return heat to the interior slowly, allowing it to cool down gradually.

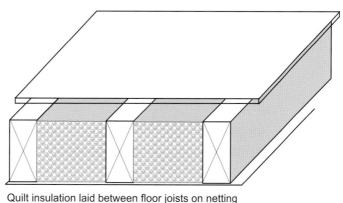

Quilt insulation laid between floor joists on netting

Figure 13.12 Improving a suspended timber floor.

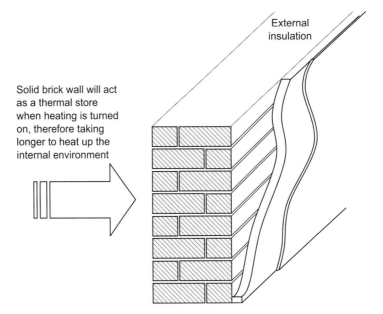

Solid brick wall will act as a thermal store when heating is turned on, therefore taking longer to heat up the internal environment

External insulation

Figure 13.13   Thermal store.

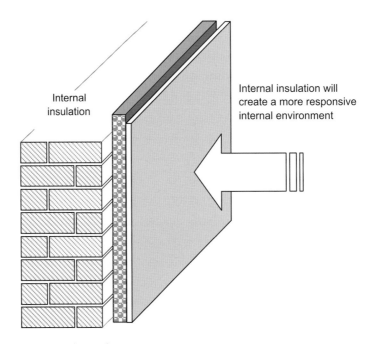

Internal insulation

Internal insulation will create a more responsive internal environment

Figure 13.14   Responsive environment.

Technically, this ability of a building to heat up and cool down is known as its thermal response time. Timber framed buildings and buildings of traditional brick and block construction behave entirely differently and with today's requirement for heating systems to be fully controllable (i.e. with thermostats that control the maximum and minimum temperatures of the installation and programmers that switch the system on and off at certain times of day and night) timber framed buildings with their quick response times can mean that it will be more efficient to heat such a building. Alternatively, in an old stone-built country house, although the response time will be slow, it means that such buildings usually stay cooler during a hot snap in summer and warmer during a cold snap in winter than an equivalent sized modern timber framed house.

Energy saving light bulbs

These bulbs, or more accurately compact fluorescent lamps (CFLs), have been used for a number of years and whilst some people have refused to use them because they can be unsightly, more modern CFLs appear very similar to traditional lamps (bulbs). CFLs are more expensive to purchase than traditional lamps. However, as they last longer this cost is offset as they do not have to be replaced so often. In recent years the cost of CFLs has come down and there has been a corresponding increase in the cost of traditional tungsten filament bulbs as we are encouraged to stop using them. They have the disadvantage that they cannot be dimmed with a dimmer switch.

CFLs are more energy efficient as they require only approximately one fifth of the energy a traditional lamp would require for the same amount of light output. Therefore, if all the lamps in a home were replaced with CFLs, the cost of lighting the home could potentially be reduced by four fifths (i.e. saving 80%). To put this into context, some sources estimate that lighting represents about 10% of the annual usage of electricity in the average dwelling. Most of this will be at peak times when electricity costs are highest, so real savings can still be made by changing to energy efficient lamps.

As well as CFLs, some spot lights can be replaced with LED lamps (or bulbs) which would also return similar improvements on energy saving.

In newly built homes the Building Regulations require that at least 1 in 4 fixed light fittings must be low energy and any security lights fixed outside a dwelling must be connected to a light detector, so that they switch off automatically during daylight hours. During night time they should be fitted with an infrared or similar movement detector, so that they only switch on when needed, and they should have a time switch to turn them off after a pre-determined interval.

Thermal upgrading of elements and changing lamps are not the only improvements that can be made. The property could also install a new A rated energy efficient condensing boiler (Chapter 12). This boiler could see a 30% efficiency increase from a standard boiler.

Installing the boiler is only part of the way to having an efficient heating system. The inclusion of suitable controls such as TRVs (thermostatic radiator valves), a room thermostat and a programmable timer will help reap the rewards of a much improved system.

It can therefore be seen that the overall thermal performance of the whole property can be improved by everyday non-complex systems or methods. Monetary savings will continue to be made, carbon emissions will be reduced and a more controllable and comfortable internal environment can be maintained.

In conclusion, the measures that are most cost effective in terms of pay-back periods (i.e. the number of years in which the cost of the work can be recouped through reductions in energy bills) can be ranked as follows:

(1) Replacement of GLS tungsten filament and tungsten halogen lamps with low-energy lamps.
(2) Roof space (loft) insulation.
(3) Cavity wall insulation.
(4) Ground floor insulation – suspended floors.
(5) External wall insulation.
(6) Replacement windows and doors.
(7) Replacement boilers and heating installations.
(8) Ground floor insulation solid floors.
(9) Internal insulation.

This is only an approximate ranking as a lot will depend on whether or not grants towards the cost of the works can be obtained. It also takes into account the disruption and difficulty of carrying out the works.

Photovoltaic panels

Photovoltaic (PV) panels are becoming much more common largely not only because of the occupant saving money on their electricity bills but also because of the opportunity of generating an income from the UK Government's Feed in Tariff (FiT). This tariff pays the owner of the panels for every kilowatt hour of electricity generated through the panels whether the generated electricity is used by the occupant or not. If the electricity is not used it is then exported back into the National Grid and an additional payment can be received from the Government.

The PV panels use energy from the sun to generate electricity. This energy is daylight, direct sunlight is not required. The panels generate a DC current, so an inverter must also be installed to convert the DC to an AC current in order that it can be fed back into the home's electrical system and used by the occupants.

The installation of the system is not a complex science but competent installers are essential to complete the electrical work and ensure an efficient system.

Several factors exist that affect the efficiency of the system, such as the type of panel (e.g. polycrystalline and monocrystalline), the length of cable runs, the orientation of the building, the pitch of the roof and the minimising of any potential shading of the panel array.

PV systems can be expensive to install but they show a guaranteed return. However, numerous companies offer to install the system free of charge. In this manner you will benefit from the free electricity generated by the panels but the installing company benefits from the FiT.

Solar water heating

Solar water heating is also becoming more common in the United Kingdom owing to the FiT (which differs slightly from PV systems) and owing to improvements in the efficiency of the systems. Natural daylight is used to heat water in the panels (or tubes) which is then used to heat the water in a cylinder.

During summer time the water can be hot enough to use without additional heating. However, during colder months additional heating will be required by the boiler system but the temperature of the water will be much higher than normal, so the panels act as a pre-heat to the existing system, which reduces the amount of energy required to heat the water. Hence, all round energy savings can be made.

Other renewable energy systems exist that are not covered in this book including:

- Wind turbines,
- Air source heat pumps and
- Ground source heat pumps.

There are many specialist companies that install these systems. A few web sites are listed at the end of this chapter.

THE GREEN DEAL

The Green Deal is the UK Coalition Government's flagship policy for improving the energy efficiency of buildings in Great Britain. It is a new market framework. It is based on a key principle that some energy efficiency related changes to properties pay for themselves, in effect, through the resulting savings on fuel bills. The Green Deal will create a new financing mechanism to allow a range of energy efficiency measures, such as loft insulation or heating controls, to be installed in people's homes and businesses at no upfront cost. The Green Deal will be available from October 2012. From 2012, people will be able to access up to £20 000 upfront to pay for energy efficiency work, repaying the costs through savings on energy bills.

For dwellings the measures that will probably qualify for inclusion in the Green Deal include the following:

Heating, ventilation and air conditioning	Condensing boilers, heating controls, under-floor heating, heat recovery systems, flue gas recovery devices
Building fabric	Cavity wall insulation, loft insulation, flat roof insulation, internal wall insulation, external wall insulation, draught proofing, floor insulation, heating system insulation (cylinder, pipes), energy efficient glazing and doors
Lighting	Lighting fittings, lighting controls
Water heating	Innovative hot water systems, water efficient taps and showers
Microgeneration	Ground and air source heat pumps, solar thermal, solar PV, biomass boilers, micro-CHP (combined heat and power)

Products installed under the Green Deal must be safe, reliable and capable of performing as intended. Products must meet existing minimum health and safety and performance standards set out in European and domestic legislation, including building regulations. For many measures, robust standards already exist. The Government is also reviewing the existing landscape of certification bodies, and the nature of warranties and guarantees that exist, and it is likely that the work will have to be carried out by members of Competent Person Schemes, members of Trustmark Schemes or similar quality control bodies although this is still under discussion.

BUILDING REGULATIONS AND APPROVED DOCUMENT GUIDANCE

The thermal upgrading works described in this chapter can be broadly divided into two main sections:

(1) Work that affects the fabric of the building (roofs, walls, floors, windows and doors).
(2) Work that affects the services installations (boilers and heating installations, other heating appliances, electrical installations, plumbing installations for hot water).

The Building Regulations that affect the building fabric are dealt with principally in Chapters 5, 6 and 8 of this book and the regulations that affect services are mainly dealt with in Chapter 10.

Much of the thermal upgrading work described in this chapter may be carried out by contractors who are members of Competent Person Schemes and the work they can carry out is discussed fully in Chapter 2. The advantages of using such a contractor are that they

- Must be suitably qualified and experienced to carry out the work for which they are registered;
- Are able to self-certify their work as being in compliance with the Building Regulations;
- Are governed by a code of practice which covers all aspects of the service they offer to customers;
- Must guarantee their work;
- Must offer insurance to support their guarantee in the event that they cease to trade during the period of the guarantee and
- Must provide a dispute resolution procedure in the event that there is a disagreement between them and their customer.

To belong to a Competent Person Scheme a company must jump various hurdles placed by the Scheme Operator, including compliance with the above listed requirements, demonstration of financial viability and trading record. For these purposes checks are made with local trading standards officers to make sure that the applicant does not have a bad record as a 'cowboy' builder.

FURTHER INFORMATION

References used in this chapter and further information can be obtained from

www.markgroup.co.uk
www.thinsulex.co.uk/
www.triiso.co.uk/
www.fitariffs.co.uk
www.sto.co.uk
www.wbs-ltd.co.uk
www.decc.gov.uk

14 Conversions

Questions addressed in this chapter:

How else can I extend my home?
What issues are associated with each conversion?
What design issues do I need to consider?
What construction issues do I need to consider?
What Building Regulations apply to these conversions?

WHAT TYPE OF CONVERSION

Much of this book is concerned with extending your home by adding extra rooms at ground and/or first floor level. There are a number of other ways to extend the living space in your property without increasing the overall footprint, depending on the nature of your existing layout and construction. These include the following:

- Loft conversions
- Garage conversions
- Basement conversions

LOFT CONVERSIONS

Converting your existing loft space into a habitable area is a common way of creating an extra bedroom. Loft conversions are often, however, fraught with many technical and regulatory issues that can make these projects more complex than first thought and, therefore, often more expensive.

In Case Study 2 at the end of this book, an example has been provided of a relatively simple loft conversion in order to set out some of the issues in greater detail. Our recommendation to anyone considering such a conversion would be to seek advice from someone who specialises in this type of work, be it a designer or contractor. We would caution, however, that you consider some of the points raised in Chapter 3 regarding the appointment of a specialist.

Extending and Improving Your Home: An Introduction, First Edition. M.J. Billington and C. Gibbs.
© 2012 M. J. Billington and C. Gibbs. Published 2012 by Blackwell Publishing Ltd.

GARAGE CONVERSIONS

Many properties have attached or integral garages as part of their construction and these offer a quite simple option for creating that extra bedroom or sitting room for growing teenagers.

Two possible scenarios are mentioned here:

- **Converting an attached garage:** This means that the garage (which is, of course, single storey) is simply built alongside the house and shares a wall in common with it. Therefore, three of the garage walls will be just a half brick (about 105–115 mm) thick and will contain one brick thick piers at about 3 m centres. The roof will usually be flat and covered in roofing felt. Often there will be no conventional foundations under the walls, but they will be built off a concrete slab with a thickened edge (and if you are very fortunate, the slab will contain steel reinforcing bars). Attempting to convert such a construction to a habitable room fully in compliance with the Building Regulations is usually a waste of time, effort and money. It is usually better to demolish the garage and build the kind of extension that you really want. On occasions, the authors have been asked to advise on the possibility of retaining the garage as a garage and extending the house on top of it to provide a first floor bedroom extension. This is even more difficult to do because of the complexities involved with carrying the first floor extension loadings down to the ground via the garage construction. It always involves underpinning the foundations and constructing extra supports for the walls at ground floor level. The cost of the underpinning alone makes it nonsensical to attempt this kind of conversion and by the time you have thickened the garage walls to take the additional loads, there is not enough room to park a decent sized car in the remainder of the garage!
- **Converting an integral garage:** This makes a lot of sense, especially if the need for an extra habitable room outweighs the need for a garage, and, let us face it, many garages are full of anything but cars! If the garden is large enough to build a separate detached garage, then you can have the best of both worlds.

Consider the conversion of an integral garage. This will normally share three walls and a ceiling with the existing house and will often have an interconnecting door into the hall or kitchen. Even integral garages are not constructed to the same standards as the rest of the home as they are not classed as habitable rooms. Therefore, the majority of the work carried out will be to improve the construction to meet current Building Regulations.

Planning permission may also be required, as the front elevation of the house will be changed when the large garage door is removed to make way for a new external wall, which might be of cavity masonry or timber frame construction incorporating a window. Most of the construction and regulatory details for the new external wall have been covered in Chapter 8 and for the new floor in Chapter 7. Works associated with garage conversions and a few issues that arise in this sort of conversion are discussed in the following sections.

Foundation construction

Sometimes the garage will have a foundation that runs across the garage door opening and which can be used to support a new external wall. The only way to establish if there is a foundation is to dig a test hole to find out its depth and width and general constructional form. More often than not, there will be no foundation, just a continuation of the garage floor slab to the outside paving of the drive. If this is the case, it will be unsuitable to take the load of the new external wall and a proper strip foundation (Chapter 4) will have to be constructed under the edge of the slab. If the garage door opening is limited in width (for example, it may be a single garage door about 2.3 m wide), it may be possible to install two suitable concrete lintels across the opening supported on the existing foundations. Again, this shows the worth of a proper site survey and investigation. Without it, you are just guessing.

Floor construction

Garage floors are normally 100–225 mm lower than the internal floor level (since they line up with outside ground level to give level access for the car) and so this provides the necessary depth to include a new timber floor (if 225 mm is available) or a new concrete floor (if less than 225 mm is available) to incorporate improved thermal performance and moisture resistance. The floor details shown in Figures 7.6 and 7.7 for a suspended timber floor may be suitable but may result in floor joists which are only 75 mm deep, so they will have to be increased in width to take the expected floor loading. Other issues that arise when building a suspended timber floor are the difficulty of installing the required amount of insulation and the need to provide fresh air ventilation under the floor. Unless the house is on a sloping site and has a good depth (>300 mm) between outside ground and ground floor level, it is usually safer, cheaper and quicker to go for a solid concrete ground floor. A suitable floor construction is illustrated in Figures 7.2–7.4.

External walls

The front elevation can be constructed with a new masonry cavity wall or in timber frame construction. If the house is reasonably modern (i.e. constructed in the past 20 years), the other perimeter walls may well be constructed to a standard that will require little or no extra insulation and are likely to be sufficiently weathertight so that no additional weatherproofing works are required. For older houses, and especially where the external walls are of solid masonry construction, it may be necessary to improve them, either by building a lightweight block wall inside the existing outer wall or by constructing a timber frame wall separated by a 50 mm cavity from the outside walls. Both of these options allow the ability to improve the thermal performance significantly and the separation prevents the movement of moisture to the inside of the building. If the external garage walls are already built in cavity masonry construction, the walls can have insulation-backed plasterboard applied to the internal surface to improve its thermal performance. All of these assumptions and options for the existing construction highlight the need for a thorough site survey and investigation, as described in Chapter 5. Typical external wall construction is described in Chapter 8. Also do not forget that any window installed will need to comply with the ventilation requirements listed in Chapter 12.

Ceiling/roof

Single-storey garages that are integral to a bungalow do not normally have the ceilings plasterboarded. Additionally, they are not usually provided with thermal insulation.

However, the lack of a ceiling allows easy access for including appropriate insulation materials and electric wiring prior to the new ceiling being fixed. Before fixing a new ceiling, the supporting structure (ceiling joists) should be checked to see whether or not they are capable of taking the additional weight.

Where the garage is integral to a two-storey house, it will probably have a bedroom, bathroom or other room above it. In this case, if the house is reasonably modern, there will be a fire-resisting ceiling between the garage and the room above. As it was originally a ceiling over a garage, it would have had 30 min fire resistance, so it would have been constructed of at least 12.5 mm of plasterboard with a 5 mm plaster skim coat or 12.5 mm of plasterboard with the joints taped and filled. Therefore, it will be more than adequate for the new room downstairs, although it may be necessary to install a mineral wool quilt to improve the sound insulation and this can only be done from above without removing the ceiling.

Services

Often the only services required for garage conversions are more electric socket outlets and improved lighting. These can be provided quite simply while other work is being undertaken so that cables can be routed appropriately. Other services such as drainage and hot/cold water may be necessary if the garage space is to be partly used as a utility room, which will then allow more space in the kitchen. Alternatively, the wish may be to include an en suite. When carrying out the site survey and investigation, watch out for the existence of gas and electricity meters and switchgear. Garages are popular places for installing such apparatus and they can be costly and difficult to relocate should this be necessary.

BASEMENT CONVERSIONS

While the majority of work associated with basement conversions is similar to loft and garage conversions (e.g. stairs, services, improving thermal performance etc.), the main difference is the ability to ensure the resistance to the passage of moisture and provide the necessary fresh air ventilation, means of escape in case of fire and the increased fire resistance needed for the floor between the basement and the rooms above. Additionally, the basement must be fit for human habitation, so it will require access to natural light and sufficient headroom. Where these are not satisfactory, the basement floor can be lowered and light wells can be constructed. Lowering a basement floor will undoubtedly require underpinning of the house foundations and is likely to be extremely expensive. Such works are often carried out in the more affluent parts of our large cities, but they should only be contemplated where the cost of the works is more than compensated for by the added value achieved by the conversion. There are several ways of addressing this issue; they are primarily specialist contractor operations. This may appear an expensive option but with such vital works as this, it is important that it is carried out correctly first time and that the work is guaranteed for the future. Basement lowering is beyond the scope of this book, as it involves specialist techniques and the services of qualified structural engineers and architects.

Waterproofing

The main two options for waterproofing basements are waterproof coatings and drained cavity systems. Waterproof coatings rely on a cement-based waterproof coating which is applied to walls and floors to protect against the lateral and vertical penetration of moisture from walls in contact with the ground and from the ground through the floor. Drained cavity systems have a profiled membrane attached to the walls and floors. This membrane is generally fixed with counter battens that allow for plasterboarding. The profile of the membrane allows moisture to drain freely down the walls to a sump that would house a pump that would eject the water once it reaches a certain level.

Some damp-proofing products can be found by using the links at the end of this chapter.

Natural light and ventilation

When creating habitable rooms (bedrooms, living rooms, studies etc.) in a basement, the conversion will be improved if it is possible to provide some form of natural light by constructing suitable light wells with windows that can be opened for purge and trickle ventilation. Additionally, the window can provide a valuable alternative means of escape in case of fire. Although building regulations do not require natural light for the habitable rooms, it might well be a planning condition that they be provided (especially if the conversion is to provide a separate flat). For kitchens, bathrooms and utility rooms, natural light is not essential; however, intermittent extract ventilation is a requirement and must be installed in accordance with the building regulations (Chapter 12).

Means of escape in case of fire

Escape from a basement fire may be particularly hazardous if an internal stair has to be used, since it will be necessary to pass through a layer of smoke and hot gases if there is a fire in the basement itself or on the ground floor of the dwelling. Therefore, any habitable room in a basement should have either

■ A protected stairway leading from the basement to a final exit or
■ An alternative escape route via a suitable door or window if it is possible to provide one. Such a window, to be acceptable as an escape route, must comply with the dimensions shown in Figure 14.1. This alternative escape route could be to an enclosed yard where the basement has access to it but certain minimum dimensions apply to the yard, as shown in Figure 14.2.

Improving the fire resistance

It is likely that if the basement has not been used for habitation before, then the ceiling (if there is one) will have a minimal standard of lining. This will need to be increased to 30 min fire resistance using the same specification described earlier for garage conversions. If the basement is being converted for separate occupation (as a basement flat, for example), then the fire resistance of the floor between the basement and the floor above would need to have a fire resistance of 60 min. This is much more difficult to achieve and

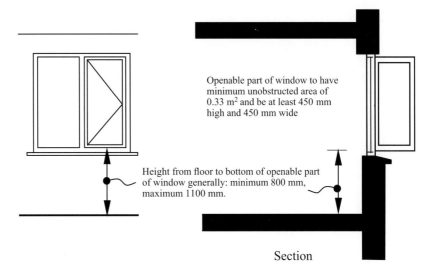

Figure 14.1   Windows for escape purposes.

puts considerable additional loading on the structure of the supporting floor, which would need to be checked by a structural engineer. Additionally, it may also be necessary to install a sound insulation quilt in the floor if either of the rooms above and below are bedrooms.

Services

The normal services of water, heating, lighting and electricity supply present no difficulties in basement conversions. This is not the case with foul drainage. If it is desired to install a new bathroom or kitchen in the basement, you may find that the sanitary fittings and sinks

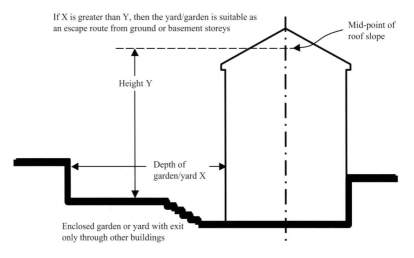

Figure 14.2   Enclosed yard or garden suitable for escape purposes in dwelling houses.

are below the level of the existing drain runs. This can be overcome by installing pumped drainage systems to lift the effluent to a level whereby it can discharge into the normal drainage system. Such systems are not cheap and come with a continuing maintenance requirement.

Approved Document H gives details of packaged pumping systems for use both inside and outside buildings as follows:

- **Inside building:** Floor mounted units available for use in basements should comply with BS EN 12050, *Wastewater lifting plants for buildings and sites: principles of construction and testing*. The pumping installation itself should be designed in accordance with BS EN 12056:2000, *Gravity drainage systems inside buildings*. Part 4: *Effluent lifting plants, layout and calculation*.
- **Outside buildings:** Package pumping installations for use outside buildings are also available. The pumping installation should be designed in accordance with BS EN 752, *Drain and sewer systems outside buildings*. Part 6: *1998 Pumping installations*.

Foul water drainage pumping installations should comply with the following:

- To allow for disruption in service, the effluent receiving chamber should be sized to contain 24-h inflow.
- For domestic use, the minimum daily discharge of foul drainage should be taken as 150 l per person per day.
- For all pumped systems, the controls should be arranged to optimise pump operation.

FURTHER INFORMATION

References used in this chapter include the following:

- Approved Document H – Drainage and Waste Disposal
- BS EN 12050, *Wastewater lifting plants for buildings and sites: principles of construction and testing*
- BS EN 12056:2000, *Gravity drainage systems inside buildings*. Part 4: *Effluent lifting plants, layout and calculation*
- BS EN 752, *Drain and sewer systems outside buildings*. Part 6: *1998 Pumping installations*

Further information can be obtained from

http://www.rentokil.co.uk
http://www.permagard.co.uk
http://basement-living.co.uk
www.fosroc.com
http://newton-membranes.co.uk/system-damp-proofing

Case Study 1   Thermal upgrading

The calculations used in this case study are to illustrate how effective thermal upgrading of a property can be. It is not intended to provide a means of calculation for Building Regulation submissions but as a comparative means of explanation. Therefore, simplified calculation methods have been used.

ELEMENTS

The existing property is a mid-terraced two storey property with the following construction:

- Roof: 50 mm insulation quilt between ceiling joists.
- External wall: 337 mm solid brickwork with lime plaster internally.
- Windows/doors single glazed.
- Suspended timber ground floor with no insulation.

Using average figures for this type of construction, the U-values for each element are calculated. (Ri is the internal surface resistance and Re is the external surface resistance.)

Roof

The U-value of the existing roof (excluding roof covering) is shown in Table A1.1.

By incorporating some additional quilt insulation on top of the existing, a new U-value for the element can be calculated (Table A1.2).

Table A1.1 Calculation of U-value of the existing roof.

Layer	Thickness (mm)	Thermal conductivity	Resistance
Ri			0.123
Plasterboard	13	0.21	0.062
Insulation	50	0.05	1.000
Re			0.055
	R =		1.240 m^{2}K/W
	Existing U = 1/R =		0.81 W/m^{2}K

Extending and Improving Your Home: An Introduction, First Edition. M.J. Billington and C. Gibbs.
© 2012 M. J. Billington and C. Gibbs. Published 2012 by Blackwell Publishing Ltd.

Table A1.2 Calculation of U-value of the roof with additional insulation.

Layer	Thickness (mm)	Thermal conductivity	Resistance
Ri			0.123
Plasterboard	13	0.21	0.062
Insulation	50	0.05	1.000
Mineral quilt	100	0.035	2.857
Mineral quilt	150	0.035	4.286
Re			0.055
	R =		8.383 m^{2}K/W
	New U = 1/R =		0.12 W/m^{2}K
	Existing U =		0.81 W/m^{2}K

Result – increasing the thermal insulation means that the roof is more than six times better at saving energy.

External wall

The U-value of the existing external wall is shown in Table A1.3.

By incorporating external wall insulation, a new U-value for the element can be calculated (Table A1.4).

Table A1.3 Calculation of U-value of the existing external wall.

Layer	Thickness (mm)	Thermal conductivity	Resistance
Ri			0.123
Lime mortar	15	0.80	0.019
Brickwork	337	0.84	0.401
Re			0.055
	R =		0.598 m^{2}K/W
	Existing U = 1/R =		1.67 W/m^{2}K

Table A1.4 Calculation of U-value of external wall with insulation.

Layer	Thickness (mm)	Thermal conductivity	Resistance
Ri			0.123
Lime mortar	15	0.80	0.019
Brickwork	337	0.84	0.401
External insulation	75	0.035	2.143
Render	13	1.00	0.013
Re			0.055
	R =		2.754 m^{2}K/W
	New U = 1/R =		0.36 W/m^{2}K
	Exist U =		1.67 W/m^{2}K

Result – increasing the thermal insulation means that the external walls are more than four times better at saving energy.

Windows/doors

This calculation does not consider the U-value of the framing material but only looks at the effects of changing single glazing to double glazing.

The U-value of the existing single glazed windows is shown in Table A1.5.

By incorporating basic double glazed units (the framing material, e.g. PVCu, timber, aluminium, is not significant), a new U-value for the element can be calculated (Table A1.6).

Table A1.5 Calculation of U-value of the existing single glazed windows.

Layer	Thickness (mm)	Thermal conductivity	Resistance
Ri			0.123
Glass	4	1.00	0.004
Re			0.055
	R =		$0.182 \, \text{m}^2\text{K/W}$
	Existing U = 1/R =		$5.49 \, \text{W/m}^2\text{K}$

Table A1.6 Calculation of U-value of double glazed windows.

Layer	Thickness (mm)	Thermal conductivity	Resistance
Ri			0.123
Glass	4	1.00	0.004
Air space			0.180
Glass	4	1.00	0.004
Re			0.055
	R =		$0.366 \, \text{m}^2\text{K/W}$
	New U = 1/R =		$2.73 \, \text{W/m}^2\text{K}$
	Existing U =		$5.49 \, \text{W/m}^2\text{K}$

Result – simply changing the single glazing to basic double glazing means that you can halve your heat losses through the windows.

In reality, it is likely that the windows themselves would be changed, since the costs of such replacements are relatively low at present. Changing to the minimum standard allowed by building regulations would mean installing a Band C window with a U-value of $1.6 \, \text{W/m}^2\text{K}$. This would mean that the new windows would be about $3\frac{1}{2}$ times better at saving energy. The top Band A windows with a U-value of about $1.0 \, \text{W/m}^2\text{K}$ would increase the energy saving performance by about $5\frac{1}{2}$ times.

Floor

The U-value of the existing suspended timber ground floor is shown in Table A1.7.

By incorporating mineral wool insulation between the joists (assuming 125 mm deep joists), a new U-value for the element can be calculated (Table A1.8).

Table A1.7 Calculation of U-value of the existing floor.

Layer	Thickness (mm)	Thermal conductivity	Resistance
Ri			0.123
Softwood	25	0.13	0.192
Re			0.055
	R =		0.370 m^{2}K/W
	Existing U = 1/R =		2.70 W/m^{2}K

Table A1.8 Calculation of U-value of insulated floor.

Layer	Thickness (mm)	Thermal conductivity	Resistance
Ri			0.123
Softwood	25	0.13	0.192
Insulation	125	0.035	3.571
Re			0.055
	R =		3.941 m^{2}K/W
	New U = 1/R =		0.25 W/m^{2}K
	Existing U =		2.70 W/m^{2}K

Result – applying thermal insulation to the floor means that the floor is now nearly 11 times better at saving energy.

HEAT LOSSES

There are two types of heat loss from a building that make up the total heat loss. Firstly, there is the fabric heat loss, which is calculated using the area of the element, the U-value of the element and the temperature difference between the internal and external environments.

Secondly, there are the heat losses due to ventilation. This happens because when a dwelling is heated the air inside it is also heated. If this heated air then escapes through draughty windows and doors and so on, energy must be used to heat up the colder air that has replaced the former heated air. Ventilation heat losses are calculated by considering the volume of the building, the number of air changes per hour (a measure of the leakiness of the building), the specific heat capacity of air and the temperature difference between the internal and external environments.

For the simplified calculation here it is assumed that the air changes per hour will not change for this dwelling, so providing calculations for this is unnecessary.

The case study mid-terraced property has the following dimensions (Figure A1.1).

Several assumptions have been made for the purpose of these calculations:

- The rear doors/windows are the same size.
- There is no heat loss through the party walls into the adjacent properties.
- A temperature difference of 15°C exists between the internal and external environment.

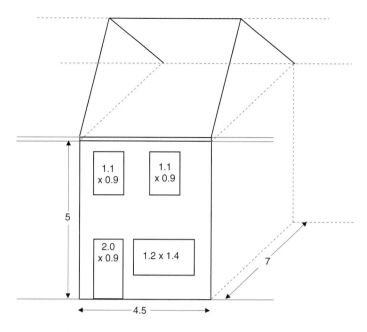

Figure A1.1   Case study property.

The heat losses through the fabric of the existing building can now be calculated using the existing U-values previously calculated (Table A1.9).

Following the thermal upgrading and new U-value calculations, the new heat losses through the fabric can be calculated (Table A1.10) to provide a comparison.

- Existing heat loss through fabric = 3412 W
- Proposed heat loss through fabric = 806 W

Table A1.9   Heat losses from existing building.

Element	U-value	Area	Temp difference	Heat loss (W)
Roof	0.81	31.5	15	383
Walls	1.67	34.08	15	854
Floor	2.7	31.5	15	1276
Windows/doors	5.49	10.92	15	899
				3412

Table A1.10   Heat losses from insulated building.

Element	U-value	Area	Temp difference	Heat loss (W)
Roof	0.12	31.5	15	57
Walls	0.36	34.08	15	184
Floor	0.25	31.5	15	118
Windows/doors	2.73	10.92	15	447
				806

This shows that by incorporating only the changes outlined above then the heat losses can be reduced by 75%.

It is worth examining the relative merits of the various measures in terms of costs and ease of application.

- The largest energy savings can be made by insulating the roof and ground floor. These measures are relatively easy and cheap to carry out.
- Applying external wall insulation is relatively expensive and difficult to do and does not result in as big a saving as the other measures already described. However, if the house had cavity masonry walls it would be a different story since this method of upgrading is easy to achieve and is also relatively cheap.
- Simply changing single glazing for double glazing is probably not cost effective, as the old draughty window frames would still exist with all their future maintenance problems. Changing, for example, to new, efficient, draught proof, uPVC windows reduces heat losses drastically in two ways – (i) by reducing the U-value and (ii) by cutting down the ventilation heat losses of the building. There is the additional benefit from uPVC windows that they do not need decorating.

The incorporation of additional energy efficient measures (e.g. compact fluorescent lamps (CFLs), an A-rated boiler and controls etc.) will assist in making even greater savings.

Case Study 2   Loft conversion

EXISTING PROPERTY

The existing property is a mid-terraced two storey property with the following construction:

- Timber cut roof with 100 mm rafters and 75 mm ceiling joists with 50 mm insulation quilt between;
- 300 mm external cavity wall.

The existing ground floor and first floor plans are shown in Figures A2.1 and A2.2, respectively, while the existing elevations are shown in Figure A2.3.

In this situation where there is a straight long flight of stairs, it is common to think about putting the new stair to the loft over this existing stair as it is 'dead' space.

PROPOSAL

The intention of this design was simply to provide another bedroom. Perhaps the word 'simple' should not be used, as even in this design there are a number of important considerations, such as:

- Can the staircase be fitted in and still comply with building regulation provisions (e.g. maximum rise of each step, minimum going of each step, maximum pitch (i.e. the slope of the stairs), headroom?
- Can timber joists for the new floor span the proposed distance or do additional steel joists have to be structurally designed?
- Can sufficient headroom be provided in the new bedroom?
- Can sufficient roof windows be inserted to provide enough light and ventilation?
- Can a suitable 'protected' route be provided in case of emergency?

The proposed ground floor, first floor and loft conversion plans are shown in Figures A2.4, A2.5 and A2.6, respectively, while the proposed new elevations are shown in Figure A2.7.

It is beneficial to have at least one roof window front and back in order that sufficient cross-flow ventilation can be provided; this will provide for a more comfortable internal environment.

The roof void areas extend into the eaves, where the maximum height would be around 700 mm. The eaves are, therefore, useless for part of the floor area but are often used to accommodate much needed storage areas.

Extending and Improving Your Home: An Introduction, First Edition. M.J. Billington and C. Gibbs.
© 2012 M. J. Billington and C. Gibbs. Published 2012 by Blackwell Publishing Ltd.

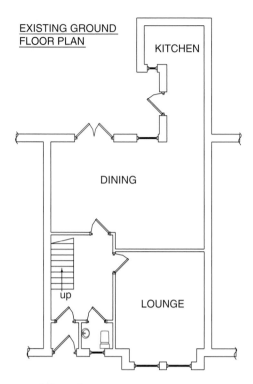

Figure A2.1   Existing ground floor plan.

The span for the floor joists is too far for a single joist span, so some form of intermediate support will be needed to shorten the distance. In this example, steel joists that will have to be structurally designed by a structural engineer, have been specified. The location of these steel joists also provides a base for the dwarf walls to be constructed. It is usually better to avoid the use of steel joists, since they are quite difficult to install at such a high level. Unfortunately, they cannot always be avoided. Where very long lengths are needed, it is possible to supply the joists in two, or three, pieces and then bolt them together. This reduces the load that has to be lifted. Figures A2.8 and A2.9 show sections through the loft.

SPECIFICATIONS

Ventilation – habitable rooms

For rapid ventilation each proposed habitable room must have an openable area of at least one-twentieth of the floor area of the room served, with some part of the opening at least 1.75 m above ground level. In addition, background ventilation must be provided having an openable area of not less than 8000 mm^2. The openable area should be controllable, secure and located to avoid undue draughts, for example a trickle ventilator.

All ventilation requirements should comply with requirements shown in Table A2.1. Bathroom and extractor fans are to have a 15 minute over-run.

EXISTING FIRST
FLOOR PLAN

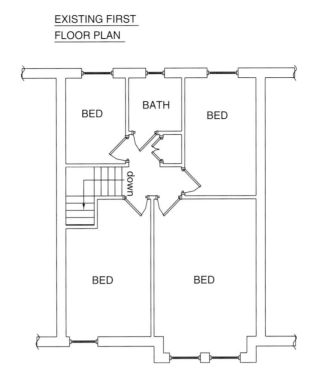

Figure A2.2 Existing first floor plan.

Windows and doors

All new roof windows to achieve a minimum U-value of $1.6\,\text{W/m}^2\text{K}$.

All existing doors to habitable rooms adjoining stairwell (escape route) to be FD20 fire doors.

Figure A2.3 Existing front and rear elevations.

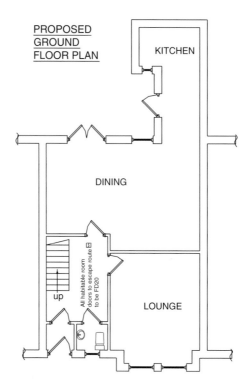

PROPOSED
GROUND
FLOOR PLAN

KITCHEN

DINING

up

All habitable room
doors to escape route
to be FD20

LOUNGE

Figure A2.4   Proposed ground floor plan.

Glazing

All new glazing to be provided with glass with low emissivity coating, such as Pilkington K glass. Safety glass to BS 6206 in all critical locations.

Electrical work

All work must be carried out by a qualified installer that must provide a signed BS 7671 electrical safety certificate.

Stairs

No more than 42 degree pitch with minimum going of 220 mm and maximum rise of 220 mm. Handrail to be provided to new stair between 900 and 1000 mm above pitch line of stair. Minimum headroom to be 1900 mm at centre of stair.

Smoke detection

An automatic smoke detection and alarm system (with a battery back-up system) based on linked smoke alarms should be installed to the standards laid down in BS 5839–6:2004 and sited where shown on the plans.

PROPOSED FIRST
FLOOR PLAN

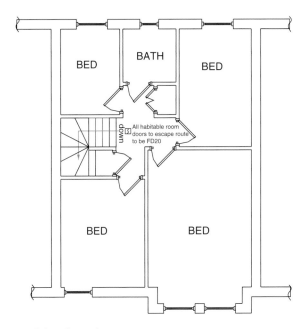

Figure A2.5 Proposed first floor plan.

Notes on section drawing

Ventilation at ridge at least equal to continuous strip 5 mm wide.

Ventilation at eaves at least equal to continuous strip 25 mm wide.

80 mm Celotex between rafters and Thinsulex multifoil insulation laid to underside of rafters and counter-battened to accept 12.5 mm vapour check plasterboard to achieve a U-value of 0.18 W/m²K.

150×38 mm rafters fixed to side of existing rafters to provide additional depth for insulation and ventilation.

28 mm tongue-and-groove chipboard flooring on 195×75 mm C16 floor joist at 400 mm centres with solid strutting maximum 2 m centre-to-centre.

100×50 mm stud panel with 100 mm Celotex insulation and 12.5 mm plasterboard to form walls and provide support and stability to roof members.

Wire mesh and 100 mm flexible slab laid over existing joists to provide fire protection and sound insulation.

New floor is independent of existing ceiling structure with joists being supported on steel joists and being placed parallel and between existing ceiling joists.

Steel joists to be designed by structural engineer.

Minimum head room to staircase preferably equal to 2000 mm but may be reduced to 1900 mm at the centre where insufficient headroom cannot be achieved.

New stud partition with 30 minute fire resistance.

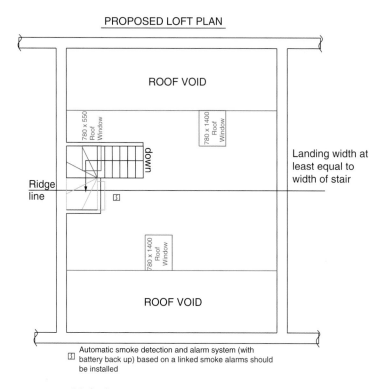

Figure A2.6   Proposed loft plan.

ADDITIONAL NOTES

More specification information would have to be provided if:

- A new dormer construction was being provided;
- An en suite bathroom is required.

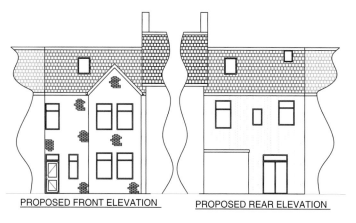

Figure A2.7   Proposed front and rear elevations.

PROPOSED SECTION

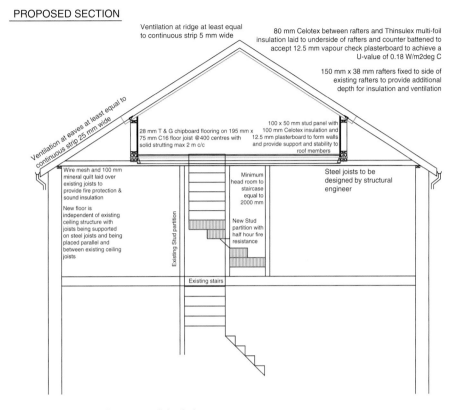

Ventilation at ridge at least equal
to continuous strip 5 mm wide

80 mm Celotex between rafters and Thinsulex multi-foil
insulation laid to underside of rafters and counter battened to
accept 12.5 mm vapour check plasterboard to achieve a
U-value of 0.18 W/m2deg C

150 mm x 38 mm rafters fixed to side of
existing rafters to provide additional
depth for insulation and ventilation

Ventilation at eaves at least equal to
continuous strip 25 mm wide

28 mm T & G chipboard flooring on 195 mm x
75 mm C16 floor joist @400 centres with
solid strutting max 2 m c/c

100 x 50 mm stud panel with
100 mm Celotex insulation and
12.5 mm plasterboard to form walls
and provide support and stability to
roof members

Wire mesh and 100 mm
mineral quilt laid over
existing joists to
provide fire protection &
sound insulation

Steel joists to be
designed by structural
engineer

Minimum
head room to
staircase
equal to
2000 mm

New floor is
independent of existing
ceiling structure with
joists being supported
on steel joists and being
placed parallel and
between existing ceiling
joists

Existing Stud partition

New Stud
partition with
half hour fire
resistance

Existing stairs

Figure A2.8 Proposed section of the loft.

ENLARGED SECTION

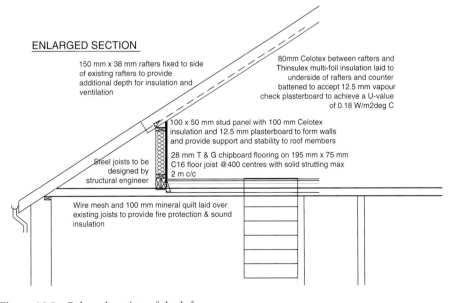

150 mm x 38 mm rafters fixed to side
of existing rafters to provide
additional depth for insulation and
ventilation

80mm Celotex between rafters and
Thinsulex multi-foil insulation laid to
underside of rafters and counter
battened to accept 12.5 mm vapour
check plasterboard to achieve a U-value
of 0.18 W/m2deg C

100 x 50 mm stud panel with 100 mm Celotex
insulation and 12.5 mm plasterboard to form walls
and provide support and stability to roof members

Steel joists to be
designed by
structural engineer

28 mm T & G chipboard flooring on 195 mm x 75 mm
C16 floor joist @400 centres with solid strutting max
2 m c/c

Wire mesh and 100 mm mineral quilt laid over
existing joists to provide fire protection & sound
insulation

Figure A2.9 Enlarged section of the loft.

Table A2.1 Ventilation requirements for the loft rooms.

Room	Rapid ventilation (e.g. opening windows)	Background ventilation	Extract/mechanical ventilation fan rates
Habitable	1/20th of floor area	5000 mm^2	N/A
Bathroom	opening window (no minimum size)	2500 mm^2	15 litres/second

Planning authorities do not look favourably on proposals that would raise the height of the ridge. Therefore, trying to maximise the headroom internally is paramount and does, of course, depend on the pitch of the roof and the existing ridge height. If additional rafters have to be placed alongside existing rafters to get extra depth for insulation purposes, this will impact on the proposed headroom. The new floor joists provide the only location where additional height can be gained. By using a 75 mm wide joist as opposed to a standard 47 mm joist the depth of the joist can be reduced and so headroom gained. This difference can be as much as 50 mm which could make all the difference. These floor joists would also be laid parallel to the existing ceiling joists but set down between them, so another possible 50 mm can be gained by adopting this practice. These are two simple, cost effective measures that could make or break a suitable proposal by achieving 100 mm additional headroom.

Important note

The suggestions given above are for information purposes only and are designed to illustrate a possible solution in the circumstances given. The authors do not accept any responsibility for use of the design in any other circumstances or when applied to any other building, since the loads, spans and dimensions will be different and will give different solutions. Interpretation of Building Regulations can vary between different building control bodies and the solutions given may not necessarily be accepted by any particular building control body.

Glossary of terms

dpc	Damp-proof course	A material that is impervious to water which is usually placed about 150 mm above outside ground level in masonry walls. Modern dpc materials are usually made from sheet plastics such as PVC.
dpm	Damp-proof membrane	A plastic sheet material placed under a concrete ground floor slab to stop water from the ground from penetrating the slab and the building.
EPC	Energy Performance Certificate	A certificate that identifies the energy rating of a property. Required by the European Energy Performance of Buildings Directive and made available whenever a building is constructed, sold or made available for rent.
EPS	Expanded polystyrene	A rigid and tough, closed-cell foam. It is usually white and made of pre-expanded polystyrene beads. Familiar uses include moulded sheets for building insulation and packing material.
TRADA	Timber Research and Development Association	This is an internationally recognised centre of excellence on the specification and use of timber and wood products.
PIR	Polyisocyanurate	Is typically produced as a foam and used as rigid thermal insulation. It is essentially an improvement on polyurethane.
PUR	Polyurethane	Is applied to the manufacture of flexible, high-resilience foam seating; rigid foam insulation panels; microcellular foam seals and gaskets.
XPS	Extruded polystyrene	Improved surface roughness and higher stiffness and reduced thermal conductivity compared to expanded polystyrene.

Extending and Improving Your Home: An Introduction, First Edition. M.J. Billington and C. Gibbs.
© 2012 M. J. Billington and C. Gibbs. Published 2012 by Blackwell Publishing Ltd.

Index

Extending and Improving Your Home: An Introduction, First Edition. M.J. Billington and C. Gibbs.
© 2012 M. J. Billington and C. Gibbs. Published 2012 by Blackwell Publishing Ltd.